The Sea's Enthrall

The Sea's Enthrall

Memoirs of an Oceanographer

Tim Parsons

© Copyright 2004, 2006 Timothy Parsons
All rights reserved. No part of this publication may be reproduced, stored in a retrieval system, or transmitted, in any form or by any means, electronic, mechanical, photocopying, recording, or otherwise, without the written prior permission of the author.

For C.V. and other professional information on Tim Parsons, visit his website at

http:www.drtimparsons.ca/

Note for Librarians: A cataloguing record for this book is available from Library and Archives Canada at www.collectionscanada.ca/amicus/index-e.html
ISBN 1-4251-1413-x

Printed in Victoria, BC, Canada. Printed on paper with minimum 30% recycled fibre. Trafford's print shop runs on "green energy" from solar, wind and other environmentally-friendly power sources.

TRAFFORD
PUBLISHING™

Offices in Canada, USA, Ireland and UK

Book sales for North America and international:
Trafford Publishing, 6E–2333 Government St.,
Victoria, BC V8T 4P4 CANADA
phone 250 383 6864 (toll-free 1 888 232 4444)
fax 250 383 6804; email to orders@trafford.com

Book sales in Europe:
Trafford Publishing (UK) Limited, 9 Park End Street, 2nd Floor
Oxford, UK OX1 1HH UNITED KINGDOM
phone +44 (0)1865 722 113 (local rate 0845 230 9601)
facsimile +44 (0)1865 722 868; info.uk@trafford.com

Order online at:
trafford.com/06-3172

10 9 8 7 6 5 4 3 2 1

For all that here on earth we dreadfull hold,
Be but as bugs to fearen babes withall,
Compared to the creatures in the seas entrall.

(Edmund Spenser, *The Faerie Queene* Book 2, Canto XII)

Acknowledgements

I wish to acknowledge and thank Anne Spencer and Valerie Green, who edited drafts of this manuscript. I am grateful to my cousin, Lorna Kingdon, who made comments and gave encouragement.

Patricia Kimber kindly copied and arranged the illustrations.

I wish to acknowledge much technical support during my research in the Pacific and Arctic Oceans and on the coast of China. Among many technicians, I would particularly like to thank Ken Stephens, Carole Bawden, Heather Dovey, Barbara Rokeby, and Janet Barwell-Clarke.

Following retirement, the University of British Columbia allowed me to retain office space for a number of years, in order to finish off some research projects. The Institute of Ocean Sciences (DFO) in Sidney, BC, provided me with an honorary research position and an office, to further continue post-retirement research and writing. I am very grateful to these two organizations for their continued support.

This book is dedicated to my family.

Table of Contents

Preface 1

PART I: "Wings on Every Wind"

I. The Beginning: Life in Ceylon 3
Early childhood–death of Father–leaving Ceylon

II. As a Boy in England 9
Growing up in Devon–my Anglo-Catholic grandfather– prepschool

III. Boarding School Years 27
Attending Christ's Hospital–learning to learn–the War in England

IV. Canada: The University Student Years 42
Emigration–farm life in Ontario–academic studies–agriculture and medicine

PART II: Seascapes

V. Learning to Swim 58
Going to sea in an oceanographic vessel–analysis of seawater–mesocosms

VI. Sailing in International Waters 66
Working for UNESCO, Paris–travels in India and Africa–other scientists

VII. Captain of My Own Boat 79
Trans-Pacific studies of the ocean–salmon enhancement on a lake

VIII. Free to Chart an Independent Course 98

1. Getting things shipshape 98
Transferring to university research–writing a textbook

2. Saving the oceans 103
Research on ocean pollution–studies in China–Aswan dam

3. Sailing against the current 118
Fisheries research–the difficulties of making academic change

4. How to train a new crew 134
Working with graduate students

5. Helping others to navigate 138
More textbooks–environmental consulting–Chile–England

6. The Northwest Passage 153
Oceanographic studies in the Arctic Ocean

7. Gumboot oceanography 159
Estuarine and intertidal studies on our coast

PART III: In Dry Dock

IX. Life in Home Port 164
Family life–role of religion

X. The Japan Prize 175

1. The origin of the Prize 175
Establishment of the Japan Prize, in 1982

2. The award 177
Purpose of the award

3. The personal side of our visit to Japan 179
Award ceremony and a visit to Kyoto

XI. A Refit for the Future 185
Early retirement and the next 30-year plan

Preface

It was the week of The Japan Prize. On the last day of celebrations, the National Theatre in Tokyo was filled with a thousand guests. They had arrived an hour before, in order to clear security, and they had been shown a vignette of the laureates' lives while waiting for Their Majesties, the Emperor and Empress. We were also waiting, in an ante-room to the theatre.

With the appropriate signal to action, the two laureates (myself and the American honoree, Dr. John Goodenough), together with our wives in formal attire, marched across a raised walkway to the music of Elgar's "Pomp and Circumstance March." The audience, dignitaries, and Their Majesties all clapped as we took our seats. The Tokyo Symphony Orchestra played the national anthem, and the presentations were made by the President of the Japan Foundation for Science and Technology. The Foundation was acknowledging my work in marine biology, or more specifically, in the field of fisheries/oceanography, and I was the first Canadian to receive this Prize. I felt very fortunate.

The Emperor concluded the presentations with a short speech about the work of the two laureates and the importance of science in the world. As we moved off-stage to the front row of the balcony, where we sat on either side of Their Majesties to listen to a concert, I looked around the sea of faces below us. I thought about the events and the many people who had led me to be at this prestigious occasion, in April of 2001. Each one of those people throughout my life, and every single event along the way, had helped me, in some way, to receive this high honor.

Therefore, while much of this book is about the oceans, it is also a reflection on the institutions of Science and Religion, family, our environment, and others I met on the long, and oft-times painful, journey that brought me to this particular day.

The winds of change directing my course were sometimes just a light breeze, at others, a gale force. How these prevailing winds affected the course I traveled is the essence of my story.

PART ONE

"Wings on Every Wind"
(George Gordon, Lord Byron, "Sonnet on Chillon," l.8, 1816)

I
The Beginning: Life in Ceylon

An old woman squatted beside her bowls of peppers and onions, spread out on the pavement. The strong, spicy smells of the meal she was cooking on her tin can stove wafted upwards into the noonday heat. Mangy dogs sniffed around the curbs, searching for scraps in the litter and dirt. A naked child played in the puddles left by the last night's rain, his brown skin glistening as he splashed himself and shouted with laughter. Bright red flowers in overhanging trees added their brilliant splashes of color to the patterns of clothing. The crowds crisscrossed the chaotic traffic, ignoring the whistles of agitated policemen. This was Colombo, Ceylon, where I was to spend the first few years of my life, a far cry from the northern countries where I would grow and find my way to a life focused on the sea.

All Saints' Day, November 1, was not a festival in Ceylon, "the teardrop of India," but for my parents in 1932, my birth in the Fraser Nursing Home, Colombo, was a modest cause for celebration. That same year, President Roosevelt was voted Man of the Year by the American people, and Joseph Stalin allowed ten million Ukrainians to starve to death through his policy of farm collectivization. The Olympic Games were held in Los Angeles, and a famous musician, John Philip Sousa, whose marching music I was to play later in school, died. Another famous musician, Glen Gould, was born in Canada.

My parents were British, and I had arrived in a colonial empire on which it was said that the sun never set. Within ten years, all that would change, as the forces of Japan swept down the entire coast of Pacific Asia in an unparalleled challenge to western colonialism The winds of change were starting to blow around the world, and as I spread my first wings, I would soon be caught up in both domestic and national events over which I had no personal control.

My first home, however, was this tropical island of less than 26,000 square miles. During its long history, it had been subject to almost continual invasions from various kings of southern India. The Arabs traded with the Singhalese in spices, particularly cinnamon, but their trade was interrupted by the arrival of the Portuguese, in 1505. In 1665, their place was taken, in turn, by the Dutch, who did much to develop a western infrastructure in the country, as well as establishing some law and order in the coastal regions; in the interior, the ruling monarch, known as the King of Kandy, remained supreme. The Dutch surrendered to the British in 1796, and the country was more or less unified under British rule until 1948, when Ceylon was granted independence under a democratic form of government.

The world that I entered was one of a ruling elite who engaged in cricket, horse racing, parties, and commerce, particularly the tea trade, which had supplanted spices in importance. The people of Ceylon were peaceful Buddhists, with some Hindus in the north. One of my earliest memories was the festival of Buddha's Tooth, which was said to have been plucked from the ashes of Lord Buddha's cremation, in 583 BC, by an attending monk. The festival was celebrated as an annual event and I well remember the many colorfully adorned elephants in the parade.

Another personal memory I have of those early days was Mother's explicit instruction regarding playing on the beach: "Timothy, if ever you see the tide go a long way out, you are to run immediately to higher ground."

Normally, the tides are very small in Ceylon, since it is close to the equator. I would probably have forgotten these instructions if it had not been that we moved to Cornwall, England, a few years later. There are very big tides on the Cornish coast and if anyone so much as mentioned that the tide was going out, I disappeared up the cliffs. It became a bit of a family joke but the interesting question is - what was the origin of my mother's instruction? The coastal areas of Ceylon were damaged by the tidal waves (tsunamis) spawned by the eruption of Krakatoa in 1883. In 1935, this was still well within the living memory of persons in Ceylon and must have been an occasional topic of conversation and concern for young mothers.

We lived first in a small bungalow, which is now the Cricket Club Restaurant, on a quiet street lined with palm trees. Here, I was to stay for four years with my parents and my older brother, Christopher, born in 1929. My father, Ernest Gordon Courtier Parsons, was a chartered accountant working for Ford, Rhodes, and Thornton. He was a tall man, and I would watch him as he left for work each morning, wearing his topee hat and cool white clothing, stark and crisp in the glare of the tropical sun. He and Mother, whom he had married in 1928, were part of the very British social scene and belonged to clubs, such as Prince's. We would go there sometimes and watch him win a skittles tournament. My mother, a trim, good-looking woman with an English rose complexion, would wear a cool, flowing dress, with a shawl and a wide brimmed hat for protection from the sun as she sat in a wicker chair, sipping a gin and tonic brought by club servants.

Our house, too, was full of servants, which was customary for the British "raj" who found the heat of Colombo, at 5°N of the equator, unconducive to housework. We had a gardener, an *ayah* (or nanny), a cook, a houseboy, and a chauffeur. My most memorable contacts were with the *ayah* and the chauffeur. The *ayah* was in charge of both my brother and me for most of the day. There may have been more than one *ayah* during my four years in Ceylon, but the one that I remember was the one who pinched—it was her form of controlling small boys, and much more discrete than a slap on the bottom. She wore white and blue saris and had bangles on her arms. When she was angry and about to pinch me, the bangles would clatter, a sound that I have continued to associate with discomfort.

The chauffeur was a small Singhalese man with a large family. He was always immaculately dressed, and his job seemed to give him status among the other servants. He was forever either polishing or driving our car, which was a Clyno.

Clynos were made in England, and in the 1930s the company was the third largest car manufacturer after Austin and Morris. As a small boy, I particularly remember this car because it had two horns; one was electrical and the other was a large rubber bulb attached to a trumpet. The chauffeur would sound both of them as we drove down the road, and the bullock carts would scatter in every direction. It gave me an infantile feeling of power to ride in Father's car! The front windshield was louvered and opened outwards, allowing a breeze to flow through to the passengers, which was pleasant in spite of the dust.

My father did not drive the car because he had had one leg amputated some years before, following a traffic accident between his motorbike and a bullock cart. This amputation would eventually lead to his death, when his artificial limb caused an abrasion of his leg and he developed blood poisoning. One shot of penicillin would probably have saved him, but that was still to be discovered. He succumbed to septicemia in December of 1935, and so I have no memories of him, save for those I mentioned earlier. When my father died, a storm cloud gathered over us.

Mother was destitute; what could she sell to raise a little money? My father had not accumulated any significant financial reserves for his young family. There were now three of us to feed, clothe, and return to England. Suddenly, we were much like refugees, having to leave everything we had behind. Mother often told me later that she did not cry as they lowered Father into the ground at the Anglican cemetery in Colombo—she said that it would have been a sign of weakness. So I came to admire her strength, of which she would need a lot more in the months ahead. Later in life, when I lost two of my own children, I had no such strength of character.

At the funeral, a Mr. and Mrs. Grieg, whom my mother knew only slightly, invited us to spend a weekend with them. They had a mountain retreat some sixty miles from Colombo, at Nueara Eliya. My mother thought that it was very kind of these brief acquaintances to make such an offer. She was suffering from deep depression after the sadness of the funeral and the realization of the bleak prospect of her penniless state. For these generous people to step forward seemed nothing less than a signal from her guardian angel that she was not forgotten.

The next day, we all bundled into the Griegs' car and headed into the country for a couple of days. The road to the mountains passed through the slums of Colombo, rice paddies, and rubber plantations, until the car started to climb into the cooler climate of the mountains. Then, like a new life, the road opened up into gorgeous shades of jungle-green, interspersed with purple and white bougainvillea blossoms, and the occasional colorful flash of a tropical bird. I remember being unhappy that night at the Griegs' house, more, I think, from the strange surroundings than from grief, as I really did not understand that Father was no longer around. This was nothing, however, to the shock that my mother was to receive on our departure the following day. Mrs. Grieg came into the bedroom and gave my mother a note.

"Here is your bill for your two nights' stay," she said. "We would like you to settle it before leaving."

Of course, I was not party to this request, or how Mother dealt with it at the time. When she told me about it later, I could understand the bewilderment and hurt that she had felt, being treated so cruelly by people she had thought of as angels.

During the next month, Mother sold the furniture, the car, and everything else she could dispose of, in order to buy three of the least expensive berths on the next Bibby Liner destined for England. She received no help from her husband's family, and this neglect of my mother and the two of us was to continue for the rest of our lives in England. Why should our paternal grandparents be saddled with the possible expense of bringing up two of their son's children? They showed no interest in us, nor did my father's two brothers. My paternal grandmother did write a letter to my mother saying she was sorry the way things had turned out, but it was of no financial assistance to have only her sympathy.

This was the stiff-upper-lip culture into which I was born; a nightmare legacy from the Victorian era so aptly described by Charles Dickens. No help for the poor. It was our bad luck, of course, and the welfare state was a few decades away. Our only recourse was to go back to my mother's father, the Reverend Frank Hawker Kingdon, who was living in a vicarage in Bridgerule, Devon, England. On January 8, 1936, our destination lay across the Arabian Sea, Suez, the Mediterranean, the Bay of Biscay, to Plymouth, England.

I have few recollections of the voyage home, except for the gully-gully man in Port Said. This Egyptian conjurer came on board in his flowing white robes and entertained us on deck. He produced a variety of animals from his clothes, from inside my shirt, and from all sorts of other impossible places. I remember laughing a lot at his superb acts, which others still practice on passengers in the Suez Canal, today.

We arrived in Plymouth about three weeks after leaving Ceylon and followed what seemed like a multitude of passengers who were put ashore in a tender, there being no docks for large liners in that port. Grandfather met the boat in his little brown Austin 7, which spluttered and backfired for

The Beginning: Life in Ceylon

fifty miles at a top speed of 30mph, as we traveled across Devon to the village of Bridgerule, which lay on the border between Devon and Cornwall. This was the first time I had met my maternal grandfather. I remember his dark suit and clerical collar. He greeted us with a big smile and gave us all a hug; this I remember particularly because of his prickly chin, where, earlier in the day, he had missed a line of stubble with his open razor.

Bridgerule was where my brother and I were to spend some of our boyhood years, in a family that included our grandfather as an important influence on our lives. My mother never married again. She told me that all she really had to live for now was the bringing up of her two boys. I am sure that my whole life would have been different with a stepfather, and I am actually very grateful to my mother for her decision.

II
As a Boy in England

Ah! happy years! once more who would not be a boy?
(Lord Byron [1788-1824] "Childe Harold's Pilgrimage")

    Grandfather was a small, energetic man, who provided most of his own food out of his vicarage vegetable garden. He spent a lot of his time walking around his parish, visiting the elderly and confined. The parish was the village of Bridgerule and the church was St. Bridget's. It was a mixture of Saxon and Norman architecture and it had a fine tower, used in olden days as a community defense for villagers when attacked by marauding Vikings. The people of his parish were, at one time, either Saxon or Celtic. My grandfather's name in Celtic, *cyning + don*, meant "king of a low hill," or "chief of the hillock." The Kingdon family can be traced back to at least 1475, when Roger Kingdon was buried at St. Hugh's Church in Quethiock, Cornwall, where there is a large brass plaque attesting to him and his family. Through many generations, the Kingdon family had always supplied a large number of clergymen to the Church of England. Grandfather's father and two of his brothers were also priests in the Anglo-Catholic branch of the Anglican communion.

Anglo-Catholics originated about thirty years before my grandfather was born, in 1860. They represented a new affirmation of Anglicanism, which became known as the Oxford Movement because it was started by some clergymen at Oxford University. It was primarily the work of John Henry Newman, who wanted to see a revival of certain Roman Catholic doctrines within the Anglican Church, as a way of revitalizing this Protestant religion. In particular, John Henry taught the apostolic succession and the integrity of the prayer book. The "apostolic succession" meant that since the time of St. Peter, only those persons ordained by the apostles and their successors had a right to teach the scriptures. This, naturally, heightened tensions amongst the Evangelists (e.g. Methodists, Baptists, and Wesleyians), who then became known, in Oxford Movement terminology, as "dissenters." Oxford Movement churches were not without their antagonists. Grandfather was always on guard against a group called Kensitites, founded by John Kensit (1853-1902). This group believed in a more personal interpretation of Christianity and were known to damage the interior of ritualistic churches. Several times, as Vicar of St. Bridget's, my grandfather would say, "I heard in the village that the Kensitites were around. I must go up and see that the church doors are locked."

Grandfather started his professional life as a teataster in London, but became involved in the Oxford Movement when his father found enough money to educate him as a clergyman. He saw Anglicanism as a compromise between Catholicism and Evangelicalism. Roman Catholicism itself was not acceptable to him because the Oxford Movement did not believe in the infallibility of the Pope. However, he strongly supported the beautification of churches, vestments, incense, and the many other rituals of the Catholic Church. A Public Worship Regulation Act, in 1874, aimed at putting down "ritualism" in the Anglican Church, was ignored by my grandfather and others of his persuasion. Within the Anglican Church, this resulted in a division into what became known as the "high" and "low" church Anglicans, but the Oxford Movement was not only concerned with ritual; its members were also concerned with everyday problems of living. This was shown in their founding of the Christian Social Union, in 1889.

The parish of Bridgerule, which my grandfather administered, was largely a poor, farming community. Most of the richer farmers in the community belonged to the evangelical churches, which made it financially difficult for Grandfather to support the beautification of St. Bridget's church. The church was in shambles at the time of my grandfather's appointment but, nevertheless, when he first arrived, he established a list of "things we needed." By the time he died, seventy-three years later, at the age of ninety-eight, while still vicar of the parish, he had accomplished all of his desiderata, including the restoration of eight bells in the tower, a beautiful reredos with a crucifix, carved in Amsterdam, the stations of the cross, and numerous figures of saints, especially St. Bridget. It was in this climate of Anglo-Catholicism that I started to grow up.

Church was the focal point of my young life and Grandfather was my father figure. On returning from Ceylon, we lived for a few years in his vicarage and I became very fond of him, not so much for his religion, but more for his support in my wanting to follow the whimsical curiosity of boyhood. He fired my imagination with stories about how Jack Frost painted everything white on clear and cold nights. We would walk in the woods around the vicarage, and he would talk about the plants and animals. He showed me how he made honey from the beehives. I watched him shoot rooks, plant vegetables, and look after endless chickens.

In other areas, I tried to observe all the religious requirements of a vicar's grandson, but I found the church services very boring and wondered more about how to increase the smoke in the incense burner than about its symbolism of prayer. Holy Week was particularly heavy, since one was not supposed to enjoy such a solemn occasion, and this often meant having to stay indoors and do something quiet. Grandfather liked to sing hymns in the evening, and the local organist would come to the vicarage to play the piano on such occasions.

Having such a powerful religious example from my grandfather at an early age left me with a terrible dichotomy. Should I seriously believe all I had been taught, or as a scientist later in life, omit much of it, but still keep alive the precious memory of my grandfather in some other way? Ultimately, I believe that I have evolved some kind of ancestral reverence which is full of love for him, but which also passes beyond his particular religious beliefs.

Grandfather's parishioners held a variety of beliefs about God. One occasion on which they showed this was the day we heard a terrific clap of thunder, while dining in the vicarage. Lightning had struck Grandfather's church and a pinnacle was knocked off the tower, crashing through the roof.

As a Boy in England

My brother saw the damage first, from the vicarage.

"Grandfather, I can only count three pinnacles on the church tower!" he exclaimed excitedly. We rushed up in the Austin 7, in disbelief. Sure enough, there was the pinnacle, all two tons of granite on the floor of the church, a big hole in the roof, and a small fire that was easily extinguished.

"It's an Act of God!" exclaimed one parishioner, who thought that not enough people were coming to church, and that this was the punishment. About a week later, Grandfather got a letter from another parishioner who said that it was an Act of God because He did not like people going to ritualistic churches. It was too bad that God was blamed either way. What was He to do? The only person who was sure it was not an Act of God was the vicar.

"If it was an Act of God," said Grandfather, "the insurance policy won't apply."

He argued that there was an occupational likelihood that tall towers could be struck by lightning, and so the insurance company paid up, although they insisted on having lightning conductors installed in future.

I especially remember the incident because there must have been some nitrous oxide generated in the church by the strike, which made me rather 'high'. I behaved like an idiot, charging up and down the church aisles, looking for something that must still be around that had all that power to knock tons of granite off a church tower. I somehow connected the "Act of God" with the need for a powerful presence. I think that it was one of those nights when I was sent to bed early because I had become overexcited, while everyone else was tired and distraught by the experience.

Mother was very attentive to her two sons, but perpetually worried about her financial situation. A small insurance policy left in England by my father was applied for maximum benefit over the years, and I believe that various relations on my mother's side of the family helped her out with finances over the next few years.

My mother was one of five children; one brother died in the First World War, at the front, and one sister was poisoned in a munitions factory during the war, but survived for a number of years afterward. A younger brother became governor of the Province of Malakal, in the Sudan, and her older sister had married a tea planter in Ceylon. My mother's mother, Jesse Kingdon, (who died before I was born) was a Freyberg—a name that came up during the Second World War, when her stepbrother was a general in the

African campaign, and later, Governor General of New Zealand. However, the name that loomed far more strongly over my grandfather and his family, was a distant relation who was a minor Cornish poet, The Rev. Robert Stephen Hawker. Apart from any distant blood relationship, Grandfather was Stephen Hawker's godson, and as children we were often recited his poems, many of which I still read with pleasure today. This is one of the better known:

> *A good sword and a trusty hand!*
> *A merry heart and true!*
> *King James' men shall understand*
> *What Cornish lads can do.*
>
> *"Trelawny he's in keep and hold,*
> *Trelawny he may die;*
> *But here's twenty thousand Cornish bold,*
> *Will know the reason why!"*

These are the first and last verses of Hawker's poem, "The Song of the Western Men," which some call the Cornish National Anthem. The words refer to the imprisonment of the Cornish Bishop Trelawny, who, in 1688, opposed James II's Declaration of Indulgence, granting tolerance to Roman Catholics in England. I think my interest in poetry throughout my life owes much to this early exposure to Hawker's works.

We were now living in Grandfather's vicarage, a large cubic building of stone with a slate roof. It was probably built early in the 1800s. There was a cellar, where Grandfather usually had a ham hanging for Christmas, along with stacks of wood for the numerous fireplaces. The many rooms included a section for the maids. Grandfather only had one live-in helper and she was the maid, cook, house cleaner, and anything else that was needed. Her name was Emily. She seemed to work all day and partly into the night. Work was her habit because I remember Grandfather telling her to go and rest, but she never would. She was a marvelous cook, and considering all she had was a wood stove in which to bake pies, roast chickens, and mix sauces, it was a wonder that she always seemed to get things right. For me, her *pièces de résistance* were her stuffing and bread sauce. The latter was a mundane sounding pap, but whatever it was she put together, I thought it was delicious.

Grandfather's other helper was David Rod, who had a speech impediment and was supposedly not too bright. He cared for the kitchen garden, which was so important to my grandfather because firstly, he loved to see plants grow, and secondly, the kitchen garden seemed to supply much of our food, especially in winter time. David also pumped the wind organ at the church on Sunday. He often fell asleep between hymns, so when Monica Bath, the organist, wanted to start a hymn with a resounding chord, there would often be only a wheezing sound emitted by the fragment of air left over in the bellows, and nothing more. Grandfather would have to yell from the pulpit, "Would someone go and wake up David?"

Besides being the organist, Monica Bath was also my nanny for a while, and took over in the vicarage when mother went away on some rare occasion. She was a lovely lady, who inspired me to form my first letters of the alphabet and showed me edible roots in the forest, which all the village people knew from one generation to another. Eventually, Monica accompanied my uncle's family to the Sudan, where she was the housekeeper in the Governor's mansion in Malakal, for a few years.

To any visitor, the most notable feature of the vicarage was the number of grandfather clocks Grandfather managed to keep running; there were five, located in various rooms and the entrance hall. In spite of his efforts to keep them all running "on time," their chimes would invariably lack synchronicity, so that the noon and midnight chimes, especially, were like a musical interlude. With much difficulty, I eventually brought one of these clocks back with me to Canada.

We had many wonderful games around the vicarage. Mother once made us a wig-wam because we were crazy about playing "Indians" in the forest nearby. We made ourselves very effective bows and arrows and hunted Grandfather's chickens and his dog. Things got a bit out of hand, however, when we started to engage bigger targets. A group of village school boys passed the vicarage every day to and from school. We took to ambushing them. At first, it seemed like good fun on both sides but then the inevitable happened, and one of the village boys was bruised a little. His mother came around the same evening with a stern warning to the vicar about controlling his hooligan grandchildren. Grandfather did not welcome this from one of his parishioners. Eventually, our rambunctious behavior led to the idea that perhaps "Bridget and the boys," as Grandfather referred to us, would be better off in a small house of our own, in the neighboring community of Bude, situated on the northern coast of Cornwall and surrounded by beautiful beaches.

Barely two years after arriving in England, we moved again, to the small house in Bude, which Mother insisted on referring to as a "bungalow"—an Indian word for a low one-storey house with lots of cross-draft to keep it cool from the heat of that country. It qualified in both respects. It was a one-storey building and the cross-drafts from the North Atlantic gales sometimes were accompanied by water under the doors and windows. Mother was forever trying to limit the internal breeze by sealing up the weather side of the house. "I hope you don't feel a draft," she would say to tea guests as the steam from the kettle was buffeted horizontally across the kitchen.

Mother had an obsession for taking walks every day, a trait that only my brother inherited. She would insist on taking the two of us out practically every day onto a windswept breakwater, regardless of weather. The breakwater protected the small harbor of Bude from the Atlantic waves, which were notorious for having caused many a ship to flounder on the Cornish coast. Some days, on such walks, we had to all hang together, or we would have been blown away. Although we were often loath to embark on such walks, in the end it was really rather an exciting experience to see and feel the awesome might of the ocean crashing against a rocky shore.

There were few luxuries that Mother could ever afford. However, one item gave her great pleasure in life, and that was the acquisition of a Biro pen. This particular pen had been invented by Ladislas Biro and his brother, in 1935, but its commercial development was thwarted by the outbreak of war, in 1939. The two Hungarian brothers set up a factory in Argentina in 1943, and after some design difficulties, the pen became a popular item in Argentina and, later, among American pilots, who found them ideal for writing at high altitudes because they did not leak, like fountain pens. The ballpoint pen, as it became known later, really hit the international scene when 10,000 went on sale at New York's Gimbles Department Store, in October of 1945. The entire stock was sold out, at $12.50 a piece, on the first day! Mother's Biro was one of the earlier models, and sheer panic would set in if she ever misplaced it in her handbag. Also, the Biro had to be sent away to be refilled, which again caused great anxiety. The progress from Grandfather's feather pens of the 1860s, which he still used into the twentieth century, to Mother's Biro, was a great domestic evolution in our family.

Just before we left the vicarage for Bude, Mother received one piece of good news. She had been trying to arrange for a public school education for my older brother, Christopher. The public school (totally private in the English system) of Christ's Hospital wrote to say they had a vacancy for her eldest boy, and so the future of his education was assured. This school was to play a pivotal role in my life also, at a later date. Having an older brother to break the ice of a British boarding school proved very helpful when Mother managed to obtain a second entrance for me.

Bude was the end of the railway line from London, so that when both my brother and I were attending boarding school in Horsham, Sussex, it was still a relatively easy 150-mile journey from home to school. The railway line also spawned a large number of holiday hotels, and while the population of Bude was probably no more than 3,000, during the holiday season it mushroomed into much larger numbers.

Although we were now living six miles from Grandfather, he continued to have a lot of influence, and to give much assistance to our family. He came down and put on his vestments when the house was ready to be lived in, and conducted a formal blessing. He also blessed my brother and me, individually. It was a very biblical moment and I have cherished it all my life.

Despite my grandfather's help, Mother continued to worry about everything, especially finances. She kept a daily account book of every penny that she spent. Some evenings she would say, "I must phone Father, I am a bit wrought up!" She also went to see the family physician about being "wrought up." He prescribed a "tonic," which seemed to be the medical solution for any kind of mental state in those days. As a result, I think that she must have drunk several gallons of Parrish's Food over her lifetime. The "tonic" had been invented by Professor Edward Parrish, in 1856, and consisted of a wide mixture of organics and inorganics, including phosphates, iron, potassium, orange flower water, and powdered cochineal, the last, no doubt, to give it its appealing sanguine color. I think that the psychological idea that she was at least taking something for her condition was the most salvific ingredient of the tonic.

Mother always did her very best at Christmas time to see that we boys had a Christmas stocking and some presents. The stocking was the most fun because whatever was in it, it was a surprise. "Mother's present" to us we usually knew about in advance because we had already tried on innumerable sweaters for size, to get the one that was finally gift-wrapped. We always spent Christmas with Grandfather at the vicarage. He was busy at this time, celebrating the Christian festival in his church, which was beautifully decorated with flowers and banners for the occasion. After the opening of presents the highlight of the day was Emily's Christmas dinner, which featured roast chicken with stuffing, vegetables, gravy, bread sauce, and a country cured ham. We especially enjoyed the plum pudding, in which silver threepenny coins were hidden. Mother always worried that I would swallow one, but I was more concerned about finding them. At this meal, it was the only time that I saw my grandfather take a drink. After dinner, he produced a bottle of port, which he proudly offered around to his guests. I always felt that Grandfather was celebrating the completion of another year at this point in his life. He just seemed so much more contented at the end of Christmas dinner than at any other time in his daily routine.

To return to Bude after Christmas, we usually took the train from Bridgerule station, which was located about two miles from the vicarage. We often had to walk to the station because during the war, petrol rationing did not allow grandfather to drive us.

For our general entertainment there was a movie theater in Bude, and we went there at least once a week, if there was something that might appeal to us. *Bambi* and *The Wizard of Oz* were among the great classics that I remember with pleasure. It always bothered me, however, that Mother bought the least expensive seats that were right at the front of the theater. All my local friends and their parents were sitting above us in a balcony, where the seats cost at least twice as much. Mother said it was a waste of money, and of course, given her financial situation, she was right. This was a situation that I was born into. It was like wearing a price tag on your collar. In class conscious Britain, one could feel the immediate superiority of boys in the balcony, versus those in the pit. It is a very difficult part of their society, from the Royal Family downwards. Many times as a youth, I heard such expressions as "Keeping to your station in life." Such grading of humans is an insult to religion, democracy, and evolution. Although I do not believe in socialism as an economic form of government, it was only after a number of strong Labour Party governments that social respect was given to a large part of the English population. Unfortunately, class consciousness among the English continues in many ways, even today. It has always had a bad effect on me, and it is a relief, now, to have escaped such a society where snobbery can be a national characteristic.

Of course, at the other end of the spectrum of social consciousness in

As a Boy in England

England is the stereotype of the delightful eccentric, who can generally talk about anything to anybody. His diet is often quite abstemious, which accounts for his being much thinner than Americans of the same age. Such people tend to dress frugally, wearing coats with leather patches at the elbows. They usually sport a funny hat and a walking stick. They are often accompanied by a dog, which is the object of much of their affection. Many of England's scientists belong to this group. Fortunately, my brother, Christopher, who is a retired naval commander, is also one of these affable Englishmen.

There were many things in Bude to engage our attention when we were young. There were miles of sandy beaches interspersed with rocky shorelines, where one could find crabs, prawns, lobsters, and conger eels, all of which I eventually learned to catch during hours there, scrambling over the rocks at low tide. Some of my time was spent with my cousin, Robert, whom I called "Bobsy." He was eight months my senior, but we were good company for one another as we roamed around the countryside.

During the war years, Bobsy's parents were in the Sudan and consequently, he spent much of his early life in England. His time was shared between an aunt, who lived about eight miles from us, and my mother and me, because he attended the preparatory school in Bude, as I did. Every weekend after school, on Friday, he left to stay with Aunt Kal. He was always miserable when he left us on Friday night, and it was not until years later that I found out why. Apparently, Aunt Kal had her own way of managing little boys, in order to avoid any disturbance. She must have seen them as being born to every kind of original sin. Bobsy said that he would often have to spend the entire weekend sitting on a chair by himself in his room. When he tried to read a book, Aunt Kal would tell him, "No ... you can't divert your attention that way. You must sit and think about your sins." I feel lucky to have escaped such a puritanical lifetime experience as Aunt Kal.

I started kindergarten in Bude, at a school called St. Petroc's. It was named after a Cornish saint, of which there are almost as many as there are in Ireland, from whence Cornwall was first Christianized. I bicycled about two miles every day, through the town and over a golf course, to reach the school. Some days, when a good southwester was blowing up the Bristol Channel, I would have a hard time staying on my bike, to say nothing of the fact that I would arrive home or at school absolutely soaked to the skin.

My first teacher was a lady called Miss Barlow. She had a severe expression that could break down into a lovely smile, if things were going right in class. Unfortunately for me, right from the beginning, I found school rather boring, compared with my exploits in the tide pools. I started to get the kind of report that read "Could do better, if only he tried."

Mother was very attentive to these discouraging signs and as my education progressed (slowly), she would tell me, "You will have to do better if you want to be somebody in this world. There will be no one to look after you when I am gone!" In spite of these strong warnings, I continued to daydream my way through most of the classes, while other "little swats" seemed to get everything right, accompanied by glowing accolades from the teacher.

Grandfather said that there were only three professions—priest, physician, and soldier. Unfortunately, none of these was of any immediate appeal to me. At one point in my life, my despairing mother found an advertisement which said that chicken sexers were in high demand. Apparently, you can only tell the sex of a newly hatched chicken immediately it is born, and failing this, your guess is as good as anyone's, until the chicks develop secondary sexual characteristics months later. Mother suggested that I might think about this as an occupation—she did not call it a profession, such as that to which my brother was heading, in naval engineering. I am afraid the suggestion went over my head, as my mind was occupied with other events.

On one occasion, Bobsy and I decided to build a boat. We did not know much about boat building at the age of eight. However, we had an idea that if we made a frame, covered it with a sheet, and then waterproofed the sheet with some kind of tar compound, we would have a boat just like the ancient cockle-boats that Bobsy said were made by early inhabitants of our land. We followed these instructions, and since the final product did not look very reliable, we decided to launch it secretly, very early in the morning. We carried the boat down to a large shoreline pool and threw it in. It floated!

"Bobsy, it floats! Jump in and I will get the paddles," I said.

Bobsy jumped in, and although he was only supposed to stand on the supporting wood frame, he landed in the middle. His two feet went straight through our improvised canvas.

"I think we're sinking!" said Bobsy.

I froze. His arms flailed about as he tried to grab the sides of the boat, which then disintegrated. Bobsy was entangled in the frame, with his two feet on the bottom of the pool, looking like some giant wading bird. Then we both collapsed, laughing. Eventually, Bobsy managed to pull himself out of the wreckage. We left the remains of our craft on the beach, but couldn't stop laughing about it all the way home. Later in the day, we went down to see what had become of the wreck. A group of smaller children was playing around it on the sandy beach, pretending to be pirates.

Other various diversions included hunting rabbits, bird nesting, and building model planes from pieces of wood. One diversion, paradoxically, was either avoiding being bullied by older boys, or engaging in some form of bullying. It seemed to be an inherent characteristic of schoolboy life. English schoolboys (and I am sure it applies to other nationalities) were very cliquish. You were either in with a group or out. At St. Petroc's I was "in" with the bicycle group which came to school every day, as opposed to being a resident in-house boarder. One day, our bicycle group decided that we must all have licenses. We made up some silly words on little scraps of paper and when any one of us wanted to challenge another we said, "Okay, show me your license!"

There was one boy, however, who was not in the group. His name was Shoebridge, and when we asked to see his license, he had none.

"Better get a license, Shoebridge, or you will be in trouble!" we told him. He was harassed for several days, until his mother came over to my mother's house and asked what kind of license she should make for her son. Mother did not know anything about it, but I just said that it had to pass with the others and left it at that. The next day, Shoebridge turned up with the most gorgeous multicolored license, carefully drawn by his mother. We wondered what he was trying to pull, making himself look better than us! Shoebridge was rapidly put down once more by someone tearing up his new license! This was not physical bullying, which also occurred, but it was probably just as damaging, psychologically. Golding's 1954 novel, *Lord of the Flies*, is, unfortunately, an only too true representation of what young boys could get up to, if left to themselves. This behavior pattern seems to come from an inherent form of biological survival and can be seen among the young of many other animals, such as wild dogs.

War broke out in 1939, and I clearly remember everyone sitting around the radio with long expressions on their faces. For many weeks following this announcement, Mother went around the town on her daily shopping tour asking, "Do you think the war will be over soon?" It seems that this was a standard question because wars were never supposed to last long. Except for the somber moods of the adults, I was not sure what all this fuss was regarding war.

On the way to school, if I ever had a couple of pennies from some kind uncle, I would stop at Mr. Smith's grocery store and buy some sweets (or candies). Mr. Smith had a special table for children to buy sweets—it was low enough so I could see what I was buying. One day, he told me that he was closing it down because the war now required ration books for sweets. At last, I understood what war was!

During the war years, the hotels in Bude were taken over by evacuees from London and other industrial areas. The boys' school of Clifton was evacuated into several large hotels. In addition to the evacuees and the public school boys, Bude became home to a large number of American soldiers, who were referred to as "dough" boys because of their lavish spending habits. As a result of this, my life suddenly became cosmopolitan, with cockney accents and the great mixture of American accents, making me more aware of the variation in life-styles.

Other signs of the war started to appear all around me. My lovely beaches were all enclosed in barbed wire and declared off limits because they were now mined. Actually, I was pretty sure that they were not, because no one had seen any mining going on. However, notices written in German and English were obviously designed as a first defense. Pill boxes and antitank obstacles of concrete also appeared, and it was said that there was a German plan to invade Cornwall and capture it as a beachhead to the rest of Britain. It was only a few months into the war when I saw my first German plane, which came over while we were out playing soccer. It dropped a couple of bombs near an anti-aircraft training camp on the hills behind us and then left. We really did not see much else in the way of German armaments, but later, when I was at boarding school, we were on a direct line of flight to London for "doodlebugs"—the pilotless V-1 bombers that Germany launched so successfully, in 1944.

One of the most thrilling episodes that occurred at this time is also featured in Canadian maritime history. On the night of August 8th, 1944, a convoy of ten ships was passing the coast of Bude under the sole escort of Canadian corvette, HMCS Regina. All was going well on this calm night, until one of the US Liberty ships, the Ezra Weston, reported hitting a mine and being severely damaged. The Regina stood by the stricken vessel and started to take her under tow when another explosion blew the Regina apart and she sank in a matter of seconds, with the loss of 30 lives. Both ships had, in fact, been torpedoed and it was probably the misinformation from the bridge of the Ezra Weston which made the Regina a sitting duck for the second attack.

For the hungry residents of Bude, this event had a beneficial impact when it was discovered that huge boxes of peanuts, chocolate bars, candies of all kinds, tinned meats, and many other kinds of groceries were floating in on the tide. Unfortunately, everything was also covered in oil from the vessels, but this did not stop us from climbing down steep cliffs to retrieve as much of the booty as we could and haul it home. It was a magnificent present to the war-starved inhabitants, despite the sorrow we felt for the loss of life on the ships.

One curious commodity that I had never seen before was a whole crate filled with boxes of what seemed to be rubber balloons. I managed to stretch one of these over my head, in order to keep the fuel oil out of my hair, and I stretched a couple more over my hands to enable me to swim better, in order to be first to claim a newly arrived box. Also, since these stretchy things obviously had great utility, I carried a whole box of them back to Mother. She wondered what on earth I was wearing when I emerged back in the garden looking like a giant frog and carrying my newly found treasures.

"Good grief!" she said. "What have you brought home now?"

She was totally disgusted with me for having arrived home with a box of French Letters (condoms). Since I had never seen such objects before, I thought I had made a brilliant discovery. I could not understand what all the fuss was about.

Another aspect of the war in Bude was the amount of unused munitions that turned up on the downlands around the town, where the military were always practicing. My cousin and I took great delight in retrieving these dangerous objects. One time, we found an unexpended magnesium flare.

"I say, Bobsy, why don't we try to get this thing going. It should really light up the place," I said.

"Crikey, yes!" replied Bobsy. "All we have to do is light a fire under it."

We took it to a field in a small valley not far from our house and lit a fire around it to see it detonate. We were disappointed when nothing happened, so we left it in the smoldering embers. About halfway home, the whole valley was suddenly illuminated in a ghostly white light.

The Sea's Enthrall: Memoirs of an Oceanographer

"It's the Second Coming," I exclaimed to Bobsy, thinking of Grandfather's prognosis of such an event.

"Don't be daft," said Bobsy. "It's the flare." The smoldering embers must have become hot enough to ignite it but, very fortunately, we were already half a mile away.

Even more dangerous was removing the explosive parts from munitions. How we avoided blowing ourselves up, I have no idea, because it was lethal work. Of course, Mother had no idea what we were doing. I remember emptying the contents of some canister into a paper envelope; we used to build an effective rocket with this and a piece of bamboo! However, on one occasion, I spilled some of the powder on the floor of the garage and, in order to hide it from Mother, I started to stomp it around with my feet. Suddenly, there was a sort of 'whoosh' under one of my feet, which knocked me over. Fortunately, being an unconfined explosion, it did no more harm than trip me up. I decided then that this was a little bit too powerful medicine to be fooling around with, and gave up disarming munitions, while probably saving my life at the same time. Taking risks because of an insatiable curiosity seems to have been an inherent part of my early life. Other boys were not so lucky, and stories abounded of children being killed by something they had found in the country.

During the war years, we had to carry gas masks with us wherever we went. These were contained in a cardboard box that hung around our necks on a string—not at all a fancy adornment and even less fancy when it came to donning them. They fitted snugly around the face and air was drawn through a filter in the front, but expelled through the rubber sides with a rude noise that mimicked severe flatulence, much to the amusement of vulgar little boys.

There were frequent rumors during the war. One most often heard was that the Germans had landed. It was believed that no official news of this was broadcast because the government did not want to panic the population. However, on one occasion, when the rumor was particularly rife, my patriotic Grandfather rang the church bells (they were to be used for no other purpose during the war). The need for this exuberance was immediately countermanded by the local home guard chief, but not before an elderly parishioner died of a heart attack on hearing THE BELLS!

During this time, Grandfather came down at least once a week to visit us and bring us a weekly supply of food to supplement our rations. He brought eggs and vegetables and also some very bitter plums in season, which he called "bullums." He advised that these should be eaten for their vitamin content. He drove his Austin 7, which lurched drunkenly as he changed gears by double-declutching on the steep hill to our house. Going down the hill in the car was actually more fun because in order to save petrol, Grandfather would start out by letting gravity do the work, without running the engine. Near the bottom of the hill, Grandfather would plunge the car into gear, which both started the motor and threw any occupants out of their seats (seatbelts not yet having been invented). You could always hear him coming or going half a mile away, with the grinding of gears and backfiring of the motor.

Grandfather was a very healthy man, and I would attribute this to his diet of fresh vegetables. However, he also believed in patent medicines, among which were Epsom and Kruschen Salts. The benefits of these salts of magnesium, sodium, and sulfate were widely advertised as being good for many ailments. That "Kruschen Feeling" was a catch phrase of the manufacturers, which was meant to imply vigorous health. Later, at school, I was to find the same curious medical philosophy applied to laxatives. All the boys in our school were dosed about once a month with a very bitter tasting laxative, cascara. The British attributed many ailments to constipation, and schools had a particular horror of this condition. When arriving at my preparatory school after an hour of biking in pouring rain, and consequently suffering from acute hypothermia, the first question that matron would ask me was, "Have you *been* today, Parsons?"

During this period, my mother decided to help in the war effort. Many years before she started to raise a family she had been schooled to be a Physical Training Instructor, and she played her part in the First World War as a member of the ladies' Air Training Corps. She took joy rides in biplanes, traveled to France, and was the "officer in charge" of a large group of women who took part in the Royal Naval, Military, and Air Force Tournament, at Olympia, London, in 1918. Now, at the beginning of the Second World War, Mother again took up teaching Physical Training (PT) for factory workers.

She had an evening class, to which I was taken because we could not afford a baby sitter. I sat at the back of the hall with many women in front of me, all facing my mother. Her exercises were accompanied by a piano player, Rosey, and together they would play and sing the tune, "Hands, knees and bumpsa-daisy, I like a bustle that bends." It was on the latter part of this refrain that all the ladies bent over, and my eyes were treated to this absolute sea of bottoms. As a rather small boy of seven, I had never realized before how very large some people were when viewed entirely from the rear. My young mind seemed to work a lot in dimensions. How long? How far? How big? How much time? I suppose it was really a matter of growing up and taking in more and more of the world around me. However, the scene at the PT classes was certainly an early introduction to the dimensions of the female form, which was not a subject that I had covered yet, in my preparatory school.

In November of 1941, my mother took me to London, where I had to write an exam for entrance to the English public school, Christ's Hospital. This school was founded by Edward VI in 1553, and was designed to make use of Greyfriars monastery, which had been vacated when Henry VIII decided that he had had enough of popes and monks. As the history books say, he "dissolved the monasteries." In 1552, Bishop Ridley of London preached a sermon, which was attended by King Edward VI. In his sermon, Bishop Ridley exhorted the rich to be kind to the poor. Edward was so impressed by the Bishop that he sent for him, and with the cooperation of the Lord Mayor of London, Sir Richard Dodds, they contrived a plan to take all the poor boys and girls off the streets of London and put them in a school to be named Christ's Hospital—and to be known as "a royal, religious, and charitable foundation." Both the boys' and the girls' schools originally occupied the Greyfriars monastery, but both were moved to the country in the early 1900s. The boys were moved to Horsham, where very beautiful monastic buildings were constructed for the containment of about 800 students. About 270 girls were moved to Hertford and housed in rather Spartan residences. According to recent reports in the Times of London, the once charity school has now become the richest public school in England, due to its many endowments.

The school had a very distinctive uniform which was, essentially, the clothing worn by Puritans in 1553. It consisted of knee britches with silver buttons, bright yellow stockings, a white shirt, and white bands which came outside a very dark blue coat. The latter was fastened by more silver buttons, and a leather girdle went around the waist outside. The clothing is well recognized in England, and Christ's Hospital boys are often referred to as "bluecoat" boys, although the "blue" is so dark as to be almost black. Interestingly, when the girls' school was combined with the boys' school in the 1980s, the girls were given the option of wearing more fashionable clothes or opting for the Puritan outfit. They voted, overwhelmingly, for the traditional uniform. Every year, the whole school paraded through London on St. Matthew's Day, and had tea with the Lord Mayor. This event preserved our historical urban roots as a city school.

To enter Christ's Hospital, I had to be of poor means, and depend upon a governor to make a cash presentation on my behalf. My mother qualified admirably as to "poor means," and my governor was a friend of the family, Mrs. Clare Spurgin, who had a lot of experience with young persons, as a Justice of the Peace. The only other requirement for entry was that the child had to pass a written exam, which was why I was now in London. Since, in 1941, I could not plead idiocy as the fault of social or genetic composition, it was beholden on me to pass the entrance exam. I remember it well, along with all the admonishments, from Grandfather and Mother, on the need to pass. If I failed, I would be allowed to write it once more, but that would have entailed the cost of another trip to London.

I remember that one of the questions on the exam (an IQ test, I suppose) was to complete the next number in the sequence 1-2, 1-2, . . .? It occurred to me that someone was having difficulty counting, so I suggested 3 as the next number after 2. Obviously I came near to failing by trying to be so helpful.

While I was busy writing wrong answers, Mother talked to several of the masters, who were both setting and marking the exams for some fifty boys on the same day. She put in some good words for me, and offered her interpretation of "he could do better" as meaning "he is a promising pupil," and either from these statements, or from some guidance from my guardian angel, I passed the exam, to my mother's utter joy and relief. So it was another wind of change, in April, 1942, that carried my brother and me on a train from Bude to Christ's Hospital, near Horsham, Sussex. I would be ten years old in November of that year.

III
Boarding School Years

'T is education forms the common mind:
Just as the twig is bent the tree 's inclined.

(Alexander Pope [1688-1744] *Essay on Man*)

My brother preceded me in school by a few years, so I had been able to learn something of boarding school life from him. He always seemed to me to be much more energetic than I was, and his avid participation in sports was a characteristic of this. I looked forward to this aspect of school life but was less certain about the rest. He dutifully escorted me all the way to Christ's Hospital, where we got off the train at the school's station. Then he walked me as far as the Preparatory School on campus, and that ended his immediate responsibilities.

The housemaster of Preparatory B was Mr. Willink. He was tall and thin and had a rather dark complexion, but his eyes seemed to twinkle from beneath bushy eyebrows. He had a friendly expression but could be severe at times, as I was to discover later. However, being torn from a caring home and placed in a monastic setting was an unsettling experience for any ten-year-old. Other boys understood that "new squits" always "blubbered" for at least a week after arrival. The fact that I did not have a uniform on arrival left me feeling even more awkward, as the one boy in 800 who was still dressed in his civvies.

"Don't worry, Parsons, you will be home again soon enough," one boy said. I wished that I could have left the next day.

Recovery from this state of mind was forced on us by the rigorous timetable of events that followed. We slept on boards covered with a straw mattress, and had a straw pillow for our heads, the latter also serving as an excellent weapon during pillow fights. There were two dormitories of about twenty-five boys each, and wake-up came from the school bell at 7 a.m. How often I was to hear the terrible clanging of that bell on a freezing cold morning, when it seemed that I had only just gone to sleep! In our house, we started the day with a cold bath, which was somehow intended to invigorate us. We marched together to the dining hall for breakfast—in fact, marching as a unit of fifty boys was the way we got around for all meals and for chapel services, which seemed almost as frequent as meals.

Catering for 800 boys during the war years must have been a nightmare. Breakfast was a huge mug of tea, some porridge with lumps in it, an egg or a boiled kipper, if we were lucky, but perhaps only jam and bread. Lunch was usually the largest meal of the day, and for this we were marched into the dining hall, accompanied by the school band. For lunch we usually had some kind of pudding with meat in it—"toad-in-the-hole"—and sometimes I think it really *was* toad. Dessert was often a suet-based pudding with a thin trickle of treacle on top. Dinners were lighter fares of bread, biscuits, cheese, and cold meat. We were usually so hungry that we ate everything, even though the fish sometimes tasted rotten, and once, when we had whale meat for a couple of days, it seemed that we were eating vulcanized rubber. Grace was said by a senior boy before and after the meal, and the whole dining hall responded to a single master, who banged a gavel on his desk, instead of issuing commands. The number of times he banged his gavel dictated whether we were to start eating or clear the tables. At the end of the meal we were dismissed with an extra loud volley.

Our formal dining was supplemented in two ways. We might receive a parcel from home containing all kinds of goodies, or, we could go to the tuck shop and spend our sixpence allowance per week to buy some extra bread, or whatever Mrs. Tickner, the proprietor, had managed to put on her shelves. Not long after my arrival at school, my dear grandfather sent me a parcel. It did not contain food as expected, but I include the incident here because inside was something of far more lasting significance.

Mail arrived in the day room of our "house" and was distributed by a senior boy. He would shout out the name of the boy so blessed with contact from the outside world, and the eager recipient would rush forward to claim his letter. All items of mail were scrutinized by other boys and this particularly applied to parcels. It was always assumed that they contained tuck, which was, naturally, to be shared with your "friends." One day, I got this parcel.

"Whatcha got there Parsons? Let's see the tuck," an older boy demanded.

The parcel was addressed to me in a rather ancient script that I recognized as my grandfather's writing. Paper and cardboard had been loosely applied, the package being secured by having been wound around by several coils of different kinds of string.

"Yeah, open it up Parsons and let's see whatcha got," said another boy.

I uncoiled the string and the wrapping paper, followed by several pieces of protective cardboard, only to disclose a pile of freshly cut snowdrops.

"Snowdrops!" they yelled. "Parsons has got snowdrops!!"

"You can't eat'em," they said, "Whatcha going to do with snowdrops, Parsons?"

I was already devastated by the contents of the package as much as anyone else, and what I was going to do with them was completely beyond me. Most of the boys quickly lost interest in my snowdrops, but one suggested, "Why don't you take them up to matron?"

Matron was the only lady in the house at that time, and was in charge of the boys' health and clothing. Her name was Miss Watts, but she was known among my colleagues as "Granite" Watts because of her general lack of sympathy to boys' complaints. However, the suggestion of approaching Matron on this issue seemed like a good idea. On the way up to Matron, I saw a note inside the flowers which said, "The snowdrops in the forest were particularly beautiful this year. I wanted to share them with you. Be a good boy. Love Grandpa." Matron was not terribly helpful. The snowdrops were in quite good condition, considering that they had traveled a hundred miles in a British mail sack. Matron had an ability to peer down her nose which she did while telling me, "I suppose I could put them in a vase."

"Yes, Matron," I said, "What a good idea. They come from my grandfather and I will tell him how much you enjoyed them."

Boarding School Years

I hurried out from Matron's presence, in case she changed her mind. I have thought about this event in my life very often. My initial reaction of total disaster on not receiving what I had expected as a boy has grown into a great love for my grandfather's appreciation of Nature and his desire to pass on his feelings to me. I can never pass snowdrops growing in many places early in the year without remembering my grandfather and how he instilled in me a love for the natural world.

Moving to a British boarding school did not do much to rouse my academic interests. During the war years, we had a number of female teachers because of the shortage of masters. The former were, however, known as "masters" because "mistress" meant something else, in what was otherwise an almost totally male-dominated society. I remember that Miss Bott was one of the first masters that I encountered in preparatory class. She gave meticulous instructions in both English language and history. My mind seemed to be unable to pay attention long enough to her instructions and when she said, finally, "Now I want everyone to write an essay on what I have been describing," I would often stare blankly at my exercise book, trying to remember a single thing that she had said. I had been too busy looking out of the window of her classroom. I had a good view of some flowerbeds where there was always a great assortment of butterflies, at least in the summer months.

I had started to collect butterflies as part of my extracurricular activities, and became so enthusiastic that I ended up with by far the largest collection in the school. However, when it came to an exhibition of hobbies, which was held annually, my display of butterflies went largely unrecognized for its entomological value. Unfortunately, I had misspelled a number of labels on the species; in particular, a butterfly known as a tortoiseshell was labeled "taught-us-shell." The exhibition adjudicator wrote on my exhibit, "The most singular feature of this exhibitor's presentation is his spelling." I was crushed! All my hours of diligently chasing down some rare specimens, climbing oak trees for the purple hairstreak, charging through meadows for the painted lady, going out at night with a solution of beer and sugar to attract hawk moths, and many other devious tactics and adventures ... all had counted for nothing, if I could not spell. This very obvious weakness in my generally poor academic standing was to plague me again, later, in senior school.

The Sea's Enthrall: Memoirs of an Oceanographer

> *Hard is the task, ill fit for youthful days,*
> *To gain from judgment's voice the meed of praise,*
> *And deep the pain that youthful souls must feel,*
> *When censure damps the glow of early zeal,*
> *The flower that dies before the eastern gale,*
> *Once promised fragrance to its native vale,*
> *And many a bud that genius hailed her own,*
> *Reproof hath blighted ere its tints were known.*

(Robert Stephen Hawker [1803-1875] "Introduction and Farewell Address")

The other hobbies that I greatly enjoyed were photography and stamp collecting. I had a box camera, and learned to develop the films myself. Apart from keeping quite a good record of my friends in school, I had one singular accomplishment with my camera. At our school, there was a large water tower positioned in an architecturally fine brick column several hundred feet high, immediately behind the dining hall. It was such a prominent landmark that it was said that German bombers used it as a navigation aid on their run up to London, which was about forty miles due north. It was, of course, totally out of bounds for boys to be inside the water tower, least of all to climb it. Unfortunately, I had this idea that the best picture of the school might be taken from the top of the tower.

How to get there proved easier than I thought. I watched a serviceman enter the tower one day, using a key which he took from on top of the doorway. Early one morning, when not much else was stirring, and having enlisted a schoolboy accomplice to keep *cave* (Latin for "beware"), I found that I could just reach the key and enter the base of the tower. Inside, I discovered a ladder that seemed to rise like Jacob's ladder, almost to heaven, and then a second ladder from a mid-point that went all the way to the top, where there was a trap door. I had only one hand to grasp the ladder. In the other I carried my box camera. Carefully, I edged my way up, step by step. This was difficult enough, but suddenly, I felt my head swimming and panic taking over. It was acrophobia—something I had never experienced before. I had never tried to climb anything so high. I was halfway up. I knew I had to move somewhere. Breathing deeply, I tried to relax, and gradually overcame my fear. I really wanted that picture! Finally, I reached the top and took a photo, which I still have, but which I never dared exhibit.

Whenever I look at that picture, I remember the overwhelming feeling of terror inside the tower. I often think that such times, though seeming unimportant, are points in our lives when choices we make contribute to the qualities that affect us in later years. Learning to rely upon oneself to overcome difficulties was a necessary survival skill at school, and I needed all the practice I could get. My efforts, however, did not result in an award-winning photograph, because all you can see is the school avenue hundreds of feet below, with two large pillars of the top of the tower covering most of the view.

Mother came to visit us at Christ's Hospital at least once a term. The school allowed visits from parents, as long as they were not too frequent. Since any number of visits less than one would not establish any frequency at all, mother felt that one visit could not be viewed as excessive. She usually arrived with a parcel of tuck, which she had scrounged and saved from her own ration book. Whatever she brought was like manna from heaven.

Mother always asked to see the house master, in order to discuss our progress. She did not really like telling me what they discussed because she knew that it might upset me. Since she was only seeing me for a couple of hours, we had to make it a happy time. The highlight was to go for tea in the tuck shop, where Mrs. Tickner served a doughy bun called a "split." This was covered with as much butter and jam as she would allow us, and was consumed with great relish. Mother usually did not have any, so that we could have a bit more. Saying good-bye to Mother at the railway station was a very sad event, but was quickly forgotten in the general pace of boarding school life.

School holidays came three times a year and were always far too short. At home from school, I continued to roam the seashore, where I was enthralled with the waves, the tide pools, the seaweed-covered rocks, the many birds that soared overhead, and others that ran along the incoming tide line. Occasionally, a much older cousin, Sinclair, would come to visit our house and I always exploited his superior knowledge of Nature in any way I could. Also, I learned to play tennis during the summer holidays, the one sport that I have continued to enjoy throughout my life.

At school, sports were something I really relished and at which I did well. I enjoyed English rugby, cricket, and swimming. In neither cricket nor rugby did I excel, although I was a moderately good team player. At swimming, I managed to make the school team in my last year and I continued to swim later, during my college years. It was the camaraderie of sports, the fun of being part of a team, win or lose, which made a lot of difference in my life.

In my second year in the Preparatory School at Christ's Hospital, Mr. Willink, the housemaster, entrusted me with the role of one of six house monitors. I was eleven years old, and it seemed only fitting, or so I thought. Halfway through the term, though, I grossly misbehaved by putting on an exhibition of bird flight, using one of my bed sheets. The time was after "lights out," but it was still a bright summer evening. While at first there was a good deal of merriment from the other boys at my antics, they suddenly fell silent. I was stopped in my tracks and turned to see Mr. Willink, who had heard the disturbance from his study below. I did not like his expression and I didn't like "six of the best" that he gave me after marching me downstairs. He also took away my position as monitor, although he reinstated me before I left the house for the upper school.

I moved there the following year. There were five grades of classes for each year, and everyone from all of the upper school houses was placed in a grade. Students in Grades A, B, and C were considered the most likely to succeed in life. Those in Grades D and E were given an education sufficient to find an occupation, but they were not being prepared for a profession. Latin and Greek were taught to Grades A, B, and C, but Grades D and E were excused, and sent to manual school to learn woodwork or metalwork, instead. Some subjects were common for all five grades, such as English Literature, History, and Science, although the latter was more heavily emphasized for the ABCs. The students in the top class of senior boys, at the age of eighteen, were known as Grecians, and they were all expected to go on to university. It is not difficult to diagnose that my steady lack of progression in most subjects placed me in the E class. I did not feel any offense at this judgment; as it turned out, it was probably fortunate because, eventually, I met the one master who seemed to understand me.

In addition to the formal academic subjects, there was also a good art program and a really excellent band and choir. I had to be tested for all three and ended up in the choir, which I really rather enjoyed. We led the singing in chapel and put on concerts once or twice a year. I was asked to join the band, and since a refusal was not really acceptable, I chose to play the easiest instrument, the bass drum. I did not play it very well, however, and on one occasion I got the whole school out of step marching into the dining hall. The monotony of the steady beat led me to experiment a bit with the pace. I was playing a steady left-right, left-right beat, when I tried the effect of slowing down. All the marchers had to take longer and longer strides until they were eventually falling over each other, or trying to change step. Among eight hundred marching students the effect was hilarious. I had to laugh as I continued my funeral pace, but this was definitely not to be tolerated. The bandmaster severely reprimanded me for my lack of effort. However, one benefit that I believe I derived from playing the bass drum was the constant exercise of my arms and shoulders. This must have helped in becoming a minor swimming champion, both at school and later in college.

Another school exercise involved the senior boys taking part in military training, for at least one afternoon per week. I could choose between the army or the air force, so I elected the latter. After a year of training, we were given the opportunity to be taken up in an airplane, the main reason why I had chosen to be in the Air Training Corps. We were taken to an airport near the coast and for safety I was harnessed into a parachute. The pressure of the straps on my shoulders and the weight of the chute heightened my anticipation. I could hardly wait to jump. I went up in an Avenger fighter, which had a pilot in front, and a seat for a navigator behind and below, separated from the pilot by a bulkhead. Communication with the pilot was via a speaking tube with a wide mouthpiece, clipped to the side of the cabin. At first, the thrill of being airborne over the English Channel, everything looking so small and far below, absorbed me, but soon my head began to swim with the violent movement of the little plane. I felt a rising in my throat. I suddenly felt very sick. I reached for the speaking tube to tell the pilot that we should go back, only to find that my predecessor in the same cabin had filled it with vomit. I struggled for another fifteen minutes against the tide of motion sickness, before the pilot mercifully returned us to the ground. This short schoolboy experience in the Air Training Corps was the only military experience that I ever had. I left England at the age of 16, too young for the two years of mandatory conscription in the British armed forces.

One aspect of boarding school life that I did not like was the discipline administered by school monitors, who were allowed to operate without supervision of the masters. One boy lost his hearing from being boxed on the ears by a monitor. Hand bones could be damaged by having knuckles rapped with the heavy end of a dining knife. There was one very short and pugnacious monitor named Crofts, who took a sadistic delight in whacking boys during physical training exercises, as they bent over to touch their toes. He would creep up from behind on a likely target, smack his bottom as hard as he could, and then exclaim, "You little slacker, thought I couldn't see you, eh?" It was all, of course, gross bullying of the type reported in *Tom Brown's School Days* a century earlier.

The discipline meted out by masters was generally more formal. For the most common, detention, one had to spend a Saturday afternoon in a classroom doing some exercise, or writing out lines (e.g. "I will listen to what I am told in future"). Running penalties were also common. Twice a week the school PT instructor (a retired army sergeant) marshaled all the penalized students to run backwards and forwards at a fast trot for half an hour or more, on an asphalt parade ground. It was a very hot experience in summer and very wet and cold in winter.

During my second year in the upper school, I came to realize that there were certain masters from whom I could learn very quickly, and others from whom I could not. This was particularly true when it came to German, which was the foreign language that most of us were required to study. I made virtually no progress for about two years, until our class came under the guidance of Dr. Masson, who was head of the language school. In one year I discovered what it was we were supposed to be learning and how to learn it. I actually passed both my written and oral leaving exams in the German language, entirely on the strength of the way Dr. Masson taught. I cannot tell what it was that was especially good about his teaching but, of course, many other students have recognized that certain teachers are inspiring and others have the reverse effect.

There was one teacher in science, however, a Mr. Kirby who had an even more profound effect on me than Dr. Masson's ability to get me through my language exam. Mr. Kirby's influence was so significant for me that it laid down the basis for a profession for the rest of my life. Among other science teachers at school, Mr. Kirby was considered a bit of an eccentric. A bachelor, he was a retired army major, but he retained all the tough characteristics of army life. He tended to bark commands, rather than converse. He was most comfortable living in a tent, which he did most of the summer, while holding student army camps. In his laboratory/classroom anything could be going on at any time of the day or night, and he was constantly experimenting with his own ideas. Long before Jacques Cousteau invented underwater breathing equipment (SCUBA), Mr. Kirby was designing all sorts of face masks and rubber lungs, and experimenting with them by taking boys down to a local lake, putting his equipment on them and watching them submerge. Due to the constant unreliability of the equipment, it is a wonder that someone did not drown. Nevertheless, he persisted in the idea because he saw it (and rightly so) as a great way to do all sorts of difficult things underwater.

Mr. Kirby kept bees and a variety of dogs. His female dogs were called Chlorine and Iodine, and, later, he had a male dog named Lobster. He made his own honey and a little mead, on the side. His laboratory garden was full of vegetables off which, like my grandfather, he seemed to live most of the time. He cooked entirely for himself, one of his favorite dishes being a jellied hogshead. Due to his very basic eating habits, I tried to avoid having meals with him, but as long as he stuck to teaching science, I was 100 percent attentive.

His scientific approach was also very basic. In the first laboratory class he asked us all why a nail becomes rusty. When we could not give any satisfactory answer, he told us to find out by next week and write it up. He asked us what happens when a piece of wood burns. We had all sorts of ideas on this because most small boys rather enjoy lighting things like camp fires. Our answers were not satisfactory, so he told us to put a small piece of wood in a test tube and heat it very slowly over a Bunsen burner. "Write down what you see happening," he said, and so the classes went on, always starting with the most basic exercise, so that science, unlike many of my other academic subjects, seemed to me to belong to the real world.

I came to use his laboratory for experiments of my own, which included keeping caterpillars, collecting plants, making my first aquarium, and doing all sorts of interesting things that seemed more like fun than studying. As far as I know, none of the other science masters offered this much opportunity to students. When it came to having to prepare for exams in Chemistry and Biology, Mr. Kirby offered a crash course in the 'essence of what you have to know'. He greatly encouraged us to read our own textbooks instead of listening to him. All of this contributed to my growing realization of what it was I wanted to do, but it did more than that!

Sometime about halfway through upper school, Dr. Van Praagh, who was head of the science school, decided to hold an exam for all Grades A to E. We were all given the identical exam, and when the results came out, I came very near to the top, among almost a hundred boys. Mr. Kirby was overjoyed, though perhaps that is the wrong expression for him, for he never showed much emotion. He did arrange for me to go and see the eminent Dr.

The Sea's Enthrall: Memoirs of an Oceanographer

Van Praagh with the suggestion that perhaps I should be moved into one of the ABC grades. I remember the salient feature of our interview with Dr. Van Praagh.

"Well Kirby, the boy seems to have turned in an unusually good paper," said the doctor. "However, he obviously can't spell. He is never going to be able to make a professional career if he can't spell."

Mr. Kirby's reply to this was, "All he has to do is get a good secretary."

This was not a sufficient answer for the head of science. I trotted back to Mr. Kirby's laboratory with my tail between my legs, but I went on learning from a great teacher. I did not need the ABC recognition among the school "swats." I was quite happy to be left to dabble, both at school and at home during the holidays, in all sorts of absorbing experiments, and to read textbooks that really meant something to me. I have never learned to spell really well, but I did teach myself to type when I was at college, and usually had a friend check anything that had to be handed in for marking. When the age of spell checkers came along, I was in clover.

There was a provision in boarding school life for occasional sickness. Although I was generally healthy, I seemed to have had an almost perpetual cold. We were allowed only two handkerchiefs per week and frequently I used to have to wash and dry mine out, overnight, on the one radiator in the dormitory. Finally, during one term, I succumbed to pneumonia and was sent to the infirmary, which was known as the "Sicker." The presiding doctor was Dr. Friend, and he took a great interest in his work, keeping meticulous records on the health of the boys, which, I believe, now form a published record. There being a shortage of any kind of drugs during the war, Dr. Friend's solution to my illness was to wrap me up in a hot poultice, which Matron stayed up all night to change at frequent intervals. Somehow, between the TLC of Matron and Dr. Friend's basic knowledge of medicine, it did the trick, and although I stayed several weeks in the Sicker, I was out of immediate danger quite quickly.

The war, begun before my arrival at school, affected my life there, as it had at home. During the period of the V-1 attacks on London, our school was in a direct line of the pilotless bombers known as "doodlebugs." Quite often, during the summer of 1944, we would hear their motorbike-sounding engines roaring overhead. If the engine stopped, we all dived for cover under our desks, or anything else that was available. We had about 10 seconds to cover as much of us as we could before the doodlebug hit the ground and exploded. The Royal Air Force was charged with trying to shoot them down before they reached London, and much of this action took place over the English Channel. However, many bombs got through, still chased by an English fighter. The problem was that the flying bombs flew faster than most fighters, so there was very little time for interception. We had one doodlebug land on the school during the night. By then, we had all been moved to underground shelters to sleep, so no one was hurt, although it shattered just about every window in the school. Our underground shelter was actually a very fortunate piece of architecture. A long underground tube (which had originally been intended for marching to the dining hall without going out on rainy days) ran the whole length of the establishment. It was not used much, but it served a very important function as a sleeping place for the entire student body and faculty during the V-1 attacks.

The school was divided into sixteen "houses," with about fifty boys in each house. I was in the same house as my brother. This was probably a mistake because we tended to see too much of each other. He seemed smarter than I was; he was two years older, and simply brighter. He had the good reports, while mine were the "will never set the Thames on fire" variety. I think that I was a continual thorn in my brother's side. During the holidays, I would find different ways to tease him, and then run away before he could catch me. I was actually seeking attention, I'm sure, but the two-and-a-half-year period that separated us was, at that age, really too much to allow for easy camaraderie. Christopher settled on a naval career. From Christ's Hospital he went to Dartmouth Naval College. Mother was especially proud to have a son in the Navy. She had worked hard to see that we both grew up in an acceptable fashion, and a naval son could be regarded as the epitome of this careful upbringing.

During the war years, my mother also contributed to the war effort by serving meals to the troops, as part of an organization known as the Church Army, the Anglican equivalent of the Salvation Army. She always chose a place to work that was near the school, so that she could come and visit my brother and me at least once a term. She was stationed at Maidstone, and New Haven on the south coast, from where the invasion forces eventually left on D-day.

Soon after the war, personal factors started to have a powerful effect on my life. As a teenager, a flood of hormones started to surge through my body, and all the things that I used to do spontaneously now seemed to require endless thought. The changes of adolescence were the beginning of a bicameral existence, when the spontaneity of youth suddenly had to be tempered by rational judgment. I could no longer run around like an idiot, pretending to be an airplane or a steamroller, as my mind chose. Now, I had to have a reason for even the slightest movement. I found adolescence extremely difficult to take. I became unutterably shy and had great difficulty sitting in railway carriages, in case any one spoke to me. At the same time, I started to notice the opposite sex. Girls, to me, had always been "cats" and were never worth a passing remark. Now, for some unexplained reason, they were looking rather interesting.

I remember one in particular, whose name was Jill Melville. I saw her a few times during the holidays, but she did not know who I was. However, in my confused boy/youth transition, I tried to impress her by riding my bike past her very fast and then, when she did not look back, riding past her again even faster. I hit the side of the pavement and fell off my bike. She didn't even look at me! She must have thought I was a real weirdo, although actually, I don't think that she noticed me at all, which was even worse.

While an acute self-consciousness had taken hold of my body in what was almost a split personality between child and adult, my academic studies, strangely, did not suffer. If anything, I became more interested in studying, perhaps because it relieved me from doing anything else where I might be the center of attention, or so I thought. The school certificate exams were coming up and after these, if a boy failed, he essentially got the boot. I wrote my exams with the masters' full expectation that I would fail. Apart from Mr. Kirby, however, I had learned some interesting facts from a history teacher, Mr. Roberts, and from a geography teacher, Mr. Carey. From these three teachers, with Dr. Masson, the foreign language teacher, I had acquired some background for the certificate exams. It was actually not enough to obtain a pass in the school certificate as set by the Universities of Oxford and Cambridge, but if we did well enough and obtained "credit" grades, instead of just passes, then we were classified by being "exempt from London matrix." This was a complicated way of saying that the University of London would accept such scholars without requiring them to write an entrance exam.

Much to the surprise of most (including myself and my mother), I managed to get the "exempt from London matriculation" designation. What was the school going to do now with a boy who had been sidetracked for an occupational trade, but who might, in fact, qualify for a profession? The following year, I was given a classification of being a "science deputy

Grecian." I could never be in the top academic category, a Grecian, because someone had omitted to teach me Latin or Greek, which was required for this designation. So I was placed in a state of limbo, in a class which collected ABC students who were on the way down, and D and E students who were occasionally found to be on the way up.

When asked what I wanted to do, I could only think of something in biology, and the only profession I knew that involved biology was farming. It was arranged that I might go to the Agricultural College of London University, but then a much better suggestion was made by the Headmaster, to go and study in Canada.

This worked out in my favor, in a strange way. There was a Canadian Pacific Railway program which encouraged British citizens to emigrate to Canada, where there was an acute need for farm workers. At the same time, the school suggested that they would make a small scholarship available to me, if I enrolled at Macdonald College, the agricultural college for McGill University in Montreal. These two events came together when I was accepted for the BSc degree at Macdonald College, and the CP railway provided me with a letter of introduction to their agricultural agent in Montreal, Mr. Creswell. The letter is dated 4 July, 1949, and continues, "Mr. Parsons will be proceeding for placement in farm employment pending admission to Macdonald College, and your usual kind services in his behalf will be greatly appreciated."

> *Oh! the eastern winds are blowing;*
> *The breezes seem to say,*
> *We are going, we are going,*
> *To North Americay.*

(Robert Stephen Hawker [1803-1875] "The Cornish Emigrant's Song" v. 1)

I was being wafted by a gentle breeze toward a totally different destiny from any I could have imagined only a few years before. I was sixteen at the time, and my mother was very apprehensive about letting me go so far away, to such a different world. To give her the credit she deserves, she eventually agreed to allow me to leave England by myself, for which I have always been grateful. My world was suddenly very exciting. On August 23, 1949, I landed in Canada.

I had come on the *Empress of France*, one of the CP Empress class of liners. The voyage had been rather miserable; not from any effects of seasickness, but from my new-found state of acute shyness. Having earlier been instilled with a certain lack of confidence in myself, I now found that my adolescence caused such an acute self-consciousness that I was barely able to leave my cabin during the entire voyage. Daily, I struggled up to the dining room and sat at a table with three other people, one of them a rather lovely girl with red hair. If the two older occupants of the dining table spoke to me, I was just able to mumble a reply, but if the beautiful woman asked me something, while giving me a disarming smile, I just felt my whole face explode in a fiery rash, and all I could muster was a stifled "yes" or "no." I felt badly about my behavior, but there was little I could do about it. My boarding school had totally failed to train me to meet this kind of situation and in particular, being a school for boys, the opposite sex was now something of a total mystery. Girls were no longer "cats," they were something else, but I was not sure what. I had to find the courage to investigate this further.

IV
Canada: The University Student Years

I was met in Montreal by Judy Johnson, the eldest daughter of a wonderful Montreal family. My mother had located them through an old school friend and she hoped that they would be a backup, in case of any collapse in my emigration plans. Judy took me back to the Johnson's Westmount house, on Roslyn Ave. Although I did not stay long with them on arrival, I was to spend the next ten Christmases at their house, and on numerous other occasions they provided me with "family" support. "Uncle" Vance Johnson was a successful insurance broker, and his wife, "Aunt" Dorothy, was a very dear lady, in whom I could confide almost as freely as in my mother. Their kindness during my early days in Canada knew no bounds.

Since I was about six weeks early for registration at Macdonald College, it was arranged that I should spend the time on a farm. The Canadian Pacific Railway representative duly arranged for me to go out to Bainsville, Ontario, to assist Mr. McVickie in his harvest. Mr. McVickie was as wide as he was tall. He had a tough demeanor, inherited from his Scottish ancestors; short on words but long on work. He met me at the bus stop and I was given a room above the garage that was part of the house. It was very hot that year and the room was stifling. When it had just become cool enough to sleep, I was woken at 6 a.m. to help with the milking, and was told to take the bull out to pasture. This would seem a simple instruction, but if you had never been close up to a bull, it was a somewhat frightening experience. Mr. McVickie handed me a rope, which passed through a nose-ring worn by a rather large Holstein bull. He also handed me a crowbar and a mallet.

"Just pound the crowbar into the ground and tie the rope to it, so as to give the bull about 20 feet of pasture all around," he said.

I started out with the bull in tow, afraid that his big head and long horns were going to butt me at any moment. He was snorting and groaning and his eyes rolled around in their sockets as if he were trying to focus on a target, but I need not have worried. Bulls are more interested in cows than people, and this one wanted to join the dairy herd in the next field. The bull's rear end kept walking towards the dairy herd while I towed the front end into the adjacent field. He walked sideways, like a crab, snorting all the time because of the tight hold I had on his nose ring. I managed to complete the task, and then noticed that Mr. McVickie, having remained at the barn door, had a grin on his face–despite my obvious apprehension. I had just passed a small test of courage administered by this tough Ontario farmer.

My other memories of the farm were of the enormous quantities of food that I ate, in order to compensate for ten hours of hard labor every day; the wonderful insects that were found in Canadian fields that I had never seen before, including the praying mantis and monarch butterflies; and Mr. McVickie's lazy son, who was always trying to find a place to sleep during the day where his father wouldn't find him and put him to work. In summary, I did not think this was what I had in mind when I chose to study biology. Although I later completed my course in agriculture at McGill University, I was to change direction in professions at least twice more, before settling down to a career.

In September, 1949, I registered for a BSc in Agriculture at Macdonald College, in St. Anne de Bellevue. The campus was well planned, with red brick buildings, weeping willow trees, spacious lawns, many playing fields, and an agricultural farm associated with the college. It was located on a river that connected the Lake of Two Mountains in the north with Lac St.

Louis in the south. This gave an area for ample boating opportunity, as I was to discover in my graduate years.

Otherwise, our accommodation was not grand. First year students, or freshmen, were assigned to some post-war huts, where four students shared a bedroom/office suite, and a toilet, with another four students next door. We slept in bunk beds, and meals were taken in a dining room on campus. In addition to the agricultural students, there were female students studying Home Economics, and others in a diploma teaching course. The total population of the campus of about 500 students was about equally divided between the two sexes.

Almost immediately on registration, I felt a tone of frivolity from among my colleagues. "It was not for knowledge that we came to college, but to raise heck all the year!" This seemed to be a popular refrain which carried through with many students for the entire four years. In the first year of registration, we freshmen were inducted into college affairs by a hazing ritual carried out by third year students, at some cost to ourselves. The indignity of having to push a peanut with your nose for the length of the entire dining room was considered a source of merriment to older students; it was a form of degradation for the freshman aimed at keeping him in place in college society. Why had I imagined that a Canadian university would have a lot of serious students going around all day discussing lectures, books, ideas, and current events? Obviously, I had it wrong, and I must take part in the rah-rah class call of "Freshmen, Freshmen, well I guess, we are in it, yes, Yes, YES!"

It was not long before the class president noticed that there were some abnormal registrants among the Class of '53—our supposed graduation year. I recall that among students who were more interested in studying than playing, there were at least two veterans and about five or six others, most of whom seemed to have come from overseas. We were hauled before the class president, one day, and told that we were making it difficult for others in our year by getting high marks in spot tests. This was causing some people to get low marks, and, horror of horrors, even to fail. We obviously had a poor class attitude and needed to take note of more extracurricular activities on campus, such as football games, basketball, dances, and parties.

Those of us who enjoyed learning did not immediately follow the president's advice, being more committed to our chosen field of study. What was interesting, however, was that over the next few years, it emerged that many of us who were getting the high marks were also those who were winning in some of the extracurricular activities. In particular, I recall that one of my academic colleagues, Don Layne, was both a leading member of the rifle club *and* editor of the campus newspaper. My own activities led to winning the campus swimming championship, playing for the college tennis team, and winning the final in our last year of intra-campus debates. Ironically, the class president did have an effect on us, not by slowing down our desire to study, but in heightening our participation in some of the more orderly activities on campus.

The first two years of a BSc in Agriculture were really the same as any other science department curriculum. Maths, chemistry, physics, and biology were all emphasized, along with English literature and a few units of optional subjects. We now had to apply ourselves to the subjects, instead of being spoon-fed by a diligent schoolmaster, and the option to study or not was clearly ours. Some subjects were taught with enough enthusiasm by the professors to heighten our interests. Biology and chemistry were the two subjects that interested me most, while physics was always a bit mystical. In fact, I was so worried about passing the physics finals that I memorized and reproduced everything that was said in lectures. I obtained my highest mark in this subject while really not understanding the finer points. Chemistry and biology were much more fun, but in answering exam questions, I tended to ramble on enthusiastically and probably lost marks for not coming to the point.

As in school days, there were certain professors who made everything sound interesting and others that tended to drone on. There seemed to be an almost inverse relationship between the droners and their scientific reputation. There was, for example, a Professor Grey, who had invented some stain for bacteria which was known as Grey's stain—obviously a research achievement—but the man was a droner. Dr. Rowles (otherwise affectionately known as "Pop" Rowles) did not seem to have any claim to fame, but his lectures in physics were an absolute delight and full of both amusement and serious science. Interestingly, he was also very dedicated to college life, and bequeathed much of his estate to the college.

The displacement from an English boy's boarding school to a co-educational Canadian college was a culture shock. It took time to adjust. I had not received any training in the subject of the opposite sex. How I should deal with the beautiful colleagues who were now ever present at most of the lectures, as well as being part of the after hours scene, was a confusing question to my mind.

I went to the Saturday night dances because that's where everyone else went. I sweated beads on the side of the dance hall, while trying to muster up the courage to ask someone on the opposite side of the room to dance. When I finally got up to dance, I did not know the steps, and so I tended to talk at a great rate about cricket, the weather, politics, money, religion, and where I came from (as if my English accent were not enough to identify my origin). Somehow, none of this led to my having a permanent girlfriend, which most of the other fellows seemed to acquire. Perhaps this was all a blessing in disguise, because I was neither to be distracted by regular dates, nor by an excessive number of basketball or football games, which were so popular among young couples.

My great friend among male colleagues, in my first year of college, was an older student named "Scotty" Patton. He had attended Gordonstoun School in the United Kingdom, and had served his National Service in the army before coming to Canada. He stood about 6 feet 5 inches, sported a black beard and, sometimes, a cowboy hat—a towering presence, but a timid person, whose great desire seemed to be to please others with his polite manners and rather humble ways. His family must have had money because he imported a rare car called a Lee Francis. It was an expensive toy with classical European lines of the 1930s; the one and only dealer in North America was in New York City, which made replacement parts hard to come by. Scotty felt about as out of place as I did during my first year of college, but we compensated for this by taking tours in his magnificent roadster, traveling around the Quebec countryside on weekends.

About halfway through the year, Scotty took a liking to the girlfriend of one of the college wardens. The warden found out about these modest overtures and planned revenge. He made his master key available to a small group of fellow students, who came into our bunkhouse in the dead of night and dumped two ash cans of water on Scotty, as he lay asleep in bed. Scotty didn't think too much of the incident, except to say that he felt that the warden had a peculiar idea of maintaining law and order among the undergraduates. Was this an extension of schoolboy bullying, or was there something more sinister in the "screw you" attitude of the warden? He had assumed that his action was *incognito,* but it was quickly common knowledge. A person with administrative power had used it to his own ends. Later on, I was reminded of this incident when I saw professional competition between individuals resulting in similar instances. It often happened that an administratively powerful person would throw a wet blanket on others' ideas or performance, in order to further his own ends.

Certain extracurricular aspects of college life were very stimulating. Every year some students put on a musical show and this often reached near-professional heights. The comedy and parody of campus life were very well done by a number of talented amateur actors and actresses. A group of West Indian students who attended college were lively and entertaining and often added a new dimension to our lives. Other interesting events included visits from Prime Minister Louis St. Laurent and Dr. Ralph Bunche (the prominent Director of UN Trusteeships), who came to address the student body on separate occasions, and this gave us some feeling of national and international relevance. In contrast, our own Dean of Agriculture, who gave at least one address per year to students and faculty, was a rather uninspiring man, long on technocracy but short on ennobling ideas.

Student "government" was represented by the Students' Council, which was an elected body of about twelve persons, taken mostly from students in their final year. The role President of the Council was the most prestigious on campus, followed by that of the representative for athletics, who always had a lot to say. I was elected to the Council in the lowly position of Coffee Shop representative. I had to liaise with the only private enterprise establishment on campus, where there were alternative foods to those in the catered dining room. Since the campus dining room was run by a dietitian, we could all keep healthy with her fare, but if we wanted something tasty, or "man's food," we went to the coffee shop. There really weren't any issues between the coffee shop entrepreneur and the students, but since the location also had a small jukebox dance hall, it was a popular place, and the owner required some voice on the Council.

In third and fourth years of undergraduate studies, a student had to select a special field, which might be agronomy, animal science, economics, or chemistry, among others. My choice was for chemistry, and this department was run by a delightful Irish professor, Dr. Common. When we got to the point of being able to understand his Irish brogue, he had a lot of sensible things to say and was a good teacher. He was one professor who had the characteristic of repeating the finer points of a lecture on another day, if he did not feel that he had been able to get his point across the first time. This was very unusual for a professor, since it seemingly indicated a weakness in his lecturing ability. We never took it that way. It was just that he was excessively concerned that his students should understand some point in chemistry and we really appreciated him for this. Later on, while I was still in college, he came skiing with a group of us, and was always entertaining company for the younger set.

In the chemical option of the agricultural degree, I started to find some solid ground on which to base a future career. Chemistry had a certainty about it that seemed to be lacking in other agricultural subjects, such as economics, even though it was the agricultural economists who usually ended up getting the best jobs. Analytical chemistry of agricultural and food materials started to mean something to me in a very practical way. This was particularly true when I was asked to analyze some butter samples for traces of margarine. In those days, it was illegal in Quebec to sell margarine, but nevertheless, it often found its way into butter, where it helped to spread the profits of some unscrupulous entrepreneur. There was a simple test for this adulterant. As part of my effort to pay my way through college, one of the professors had me analyze some samples collected by the government inspectors. It was a sort of reality check on my educational process to be finally interacting with society in this manner.

As for other ways of interacting, I recall two female students, during my undergraduate years, who were particular friends. One was a very well educated lady from Barbados, called Maureen. She had a wonderful sense of fair play, coupled with a mature outlook on life. Everything she said always seemed totally reasonable and there was never anything improper in her conversation. In a way, this suited me because I would not have known how to handle sensitive boy/girl subjects. She eventually married one of the agriculture students and, I hope, lived happily ever after, although I lost contact with her after college. The other lady, Sheila, came from a Montreal family and was a much more risqué conversationalist. She was full of laughter, a good athlete, and just a lot of fun to be with. Unfortunately, she was quite a different person at home, where I visited her for a few days, one summer. She had an overly dominant mother and this affected her personality. What I thought was going to be a fun visit turned out to be anything but, and I did not carry on with this relationship. Thus, my undergraduate years did not yield any lasting relationship with a young woman. Many of my colleagues did form permanent attachments during their college years, but since I did not, it is probable that the trivial quotation that was added to my class photo might have been deserved, but I never liked it. I am not sure where it comes from:

> *Which is the better position—bondage bought with a ring.*
> *Or a harem filled with beauties, fifty tied with a string?*

It sounded a bit sadistic tying up women with string, and I would have preferred a nobler quotation, but I suppose we must be as others see us, to some extent.

During the summer months between studies at college, I took a variety of jobs, always in some way associated with agriculture. I spent one summer on a farm near Ormstown, Quebec. Here, I learned that farmers often spread their income over a variety of products so that they are less likely to go 'bottom up' if any one specialized crop fails. I started in the farmer's saw mill in late April, making planks out of logs brought into the farm from the surrounding forests during the winter. The farmer and I, working as a team, produced quite a pile of lumber and an even bigger mound of waste wood for fireplaces. We planted corn, and milked a small herd of Holsteins.

The farmer also had some beef stock and while they were left to more or less run wild, we had to go out and fence the boundaries of the cattle range. This was one of the toughest jobs I had ever encountered because we had to repair barbed wire fences and erect posts in the most impossible terrain of peat bogs, while all the time being eaten by black flies and mosquitoes. We returned, each day, with our faces bloodied by all the bites, and our eyes swollen almost shut.

Chickens were the easiest animals on the farm to care for; a handful of grain a couple of times a day was all they needed. Much more difficult to manage were the farmer's two enormous cart horses. Putting the bit into the mouth of an animal eight feet high is no easy task, if the animal has any inclination not to cooperate. I had to learn to bring them in from pasture, install the bit, harness them, and then ride out again with them towing a variety of implements, from a hay rake to a manure spreader. The latter was a hazardous implement because as I drove along, the manure was tossed in little bits into the air, and although these were supposed to land in the field, several chunks often ended up on top of my head or down the back of my shirt. Haymaking was physically the second most strenuous occupation, but unlike fencing in the spring, there were fewer bugs to eat you in the summer, and it was a relatively clean occupation. The whole exercise of working on a farm reinforced my earlier conclusion that I did not really want to be a farmer.

In April of 1953, I received my BScAgr (with distinction). The Johnsons attended my graduation, and I went home for a month, by sea, to visit Mother and Grandfather. Twice during my four years of undergraduate studies my mother had come out to Canada to visit me, and twice I had gone home for two weeks at Christmas. One of the visits by air to the UK, in the fifties, was particularly exciting. At that time, icing of piston-driven aircraft flying at about 10,000 feet was quite common and often resulted in their disappearance. We were somewhere off the coast of Ireland when icing resulted in two of the four engines shutting down. The pilot descended several thousand feet, and then a third engine quit. We flew in an unsteady fashion, and I remember seeing a Shackleton aircraft flying beside us with a lifeboat slung underneath.

For some time we seemed to come closer and closer to the white caps. I was thrilled. Still being in my teens, I had no feeling of danger. In my naivety, and with that sense of invincibility felt by so many young people who volunteer for the front lines in times of war, this was an adventure. The adrenaline rush was more like that from riding a roller coaster, than any perceived peril. The escape door was not far from my seat, and we had already donned life jackets. Most of the other passengers seemed to be very quiet and there was certainly no sign of panic. It was just pure excitement, and not until I had a family of my own did I develop a distinct fear of flying. Eventually, the pilot managed to get the faulty engines restarted, as the warm near-surface air melted the ice that was encasing them.

My ties to my mother remained strong, despite the distance that separated us, and I never missed writing to her every Sunday, until she died, in 1976. That amounted to nearly a thousand letters. I must admit that some of them said little more than that I was still around and working hard, but at least she always knew that I was thinking about her. She seemed proud of my progress, although I know that she had given up a lot in allowing me to live so far away, so early in my life.

After my BSc, I decided to continue in agricultural chemistry, toward a MSc degree. Thus, I became one of only a few students to enter into graduate school that year. Life, at once, became much more academic and less frivolous, with respect to daily events. Among my friends who had also entered graduate school, there developed the kind of camaraderie I had missed since being at boarding school. Subjects we discussed were much closer to our work, and we spent much more leisure time together. One of our group had an old car. A friend and I bought a sailboat for a very small sum because it had been unclaimed for ten years in a boat yard. There were also innumerable bicycles to be used for local trips into town.

At about this time, I met a young lady, whom I suppose I could call my "first love," if the time we spent together and our attraction for each other was any measure. Her name was Joan, and she taught kindergarten at the local high school. She had a very pleasant figure, but most of all, she had a wonderful smile and an infectious laugh. Laughing is something I love to do but, until then, I had found life to be rather serious. With Joan, it seemed that it could be a lot of fun! I continued seeing her, even after moving to Montreal, and for some years I thought that we would make a great couple after I obtained my PhD. However, that was not to be. One day, she did not reply to my phone calls, and someone told me she had got engaged. I was very upset, and started reading all the poetry I could find, such as Keats' "La Belle Dame sans Merci," where I sought consolation from others' experiences. The pain of lost love was with me for some time.

I found that my sailboat became my second love in the summer months that followed. It was 28ft. long and had a 35ft. spar. One favorite cruise was across the Lake of Two Mountains, to Oka and back, in an afternoon. My friends and I drank a lot on the way. One awkward feature of the boat was that it had a 5ft. drop keel. This could normally be dropped and wound back up again from the center well of the boat, but one day, when the cable broke off the bottom of the keel in stormy weather, it presented a problem. I had to dive under the boat and, in one breath, reattach the frayed cable to the bottom of the keel, or we would never have made it home through some of the shallower waters. This wasn't an easy task, and if I had not had a lot of swimming experience, it would have been impossible. On another expedition, I managed to break the main spar when we went about, and a stay came loose on the windward side. We limped home with the 7 horsepower outboard engine, and I had the cost of my mistake to bear with a shipbuilder.

In school, I continued my work on the hydrolytic products of proteins towards a required thesis. These protein hydrolysates were used to treat patients who required some predigested food, and my professor, Dr. Bruce Baker, and I were trying to find new and cheaper ways of producing suitable alternatives to commercially available products. In carrying out such research, one learns much incidental knowledge about how to do experiments, how to draw conclusions, how to talk about work with others, and how to present it at scientific meetings. It was a new world of quasi-professionalism and I greatly enjoyed it. We had to pass a set number of courses at the same time as we did our research; unfortunately, I failed one of these through lack of interest in the content. Failure of a course in graduate school was anything below 65 percent and could result in dismissal. However, my professor allowed me to rewrite the exam, and I passed.

Although the subject of my research was nominally agricultural, it obviously had a medical application. It was during these studies that I started to gravitate towards research in medicine. This was particularly reinforced when I had to take one course in biochemistry, on the McGill campus in Montreal, which was given by the Dean of Graduate Studies, Professor David Thomson. He was the son of Sir Arthur Thomson, the celebrated naturalist who wrote many learned works on corals, but who was also well known for his books on the reconciliation of science and religion (e.g. *Science and Religion*, 1925). I became a good friend of his grandson, John Thomson, a contemporary of mine at McGill. In the future, John's son, Andrew Thomson, would take a course from me at the University of British Columbia. In one way or another, I was to be connected with this family throughout my life.

The Department of Biochemistry at McGill University was in the Faculty of Medicine, and Professor Thomson's lectures contained some of the most interesting science I had yet encountered. His influence on me was a significant force that sent me in a different direction. Chemical compounds, that I had previously seen as just something that reacted with something else, which one could analyze before and after the reaction, now appeared to be living ingredients of my body, where they had more than properties, they had function! This sounded like a branch of science that I would really like to pursue, rather than one that led to McVickie's bull, or black flies in the swamps of Quebec. I obtained my MSc in 1955, and published my first paper on the results of my thesis work. One's first publication is a source of considerable pride, even though its author is still a neophyte to the world of research.

I discussed my future with the head of chemistry at Macdonald College, who suggested that I should continue my studies by transferring to the Department of Biochemistry at McGill University. There, I was introduced to Professor Orville Denstedt, who agreed to take me on as a graduate student working toward my doctorate. He had some financial support for studies on silicosis, a disease of hard-rock miners that was prevalent in Quebec. It was caused by inhaling the dust fragments of quartz that occurred during the mining process. While the cause was known, the mechanism of just how the silicon nodules developed in the lungs was a subject for research. Dr. Denstedt had the idea that the element, silicon, leached from the quartz, caused a metabolic disturbance in the lungs, leading the formation of fibrous nodules, and eventually to asphyxiation of the patient.

I spent my first year trying to connect silicon with any kind of metabolic process in the body. This turned out to be a blind alley, and it seemed, to me, that although many organisms contained silicon, as an element, it was biologically inert. I turned the corner in my research when I decided to study the biochemical properties of quartz particles within whole animals (that meant rats, in those days), and the interaction of quartz with various enzyme systems of the body. I would not have chosen to study silicosis as a research topic if it had not been for the stipend that accompanied the project, allowing me to live through the lean years of graduate school in Montreal. One could not select special topics if there was no support money available. However, I was quite happy to be able to dabble in biochemical reactions, and the *pro bono publico* aspect was something of an incentive. I was feeling reasonably well settled, academically.

Although I was thrilled with my new research opportunity in biochemistry, it meant that I had to move from the countryside atmosphere of Macdonald College in St. Anne de Bellevue, to the city of Montreal. I had never lived in a city before, and I became desperately unhappy with this new environment. I can express this best in a short poem I wrote at the time:

On Leaving the Country for the City

Where once the trees in grandeur spread,
Their shady mantle overhead,
Now stands a lamp in morbid silence
Casting e'er its vile brilliance.

Where once a river in silent motion,
Flowed towards some mighty ocean,
Now the streets immobile stand,
Made of stone, poured by man.

The University Student Years

Where once a bird could circle by
To cast its voice upon the sky,
Now dust and dirt fly in its stead,
And only sirens sound over head.

Where once the air was clean and clear,
Now my eyes through smog appear;
Not that mist of my native valley,
But smoke and soot from the neighbor's alley.

Hard the change that I must bear,
That progress may serve those who care
For city living, when once I could,
Talk to Nature, and through Nature, to God.

Again, as I was to find throughout emotional periods in my life, I turned to poetry for solace. Through reading the works of others, I could find ways of understanding feelings. In writing my own poetry, I could express my inner thoughts and emotions more deeply than in prose. Though my skill was not as great as that of my poet ancestor, Robert Stephen Hawker, I felt a kinship with him at these times. While many people do not read poetry, I feel it provides a medium more satisfying than prose. For me, in the city, it conveyed my sense of alienation.

It was 1955, about a decade before the environmental movement started on this continent. I must already have been conscious that all was not quite right in the way that urban sprawl had developed and taken over so much of our green space. I am not sure how God got into the last line of the poem, but in transforming my religion from strict Anglo-Catholicism to more liberal thoughts, I have seen this world as the product of some kind of intelligence. I think of Nature as being the most incredibly clever creation possible, and it is as a presence within Nature that I imagine a Creator exists.

My colleagues in the Biochemistry Department at McGill University were a very different company from those in graduate school at Macdonald College. For the first time, I was thrown into an arena of intense competition. It was no longer a relaxed atmosphere of research to uncover some new fact; now, it was research to write the most significant paper the world had ever known, for such journals as *Nature* or *Science*. This was a climate in which I first heard the Nobel Prizes discussed. Who was getting them? What did they actually discover? A large number of recipients, it seemed, were Jewish, as were most of my colleagues, some of whom already had MD degrees, but were undertaking PhDs, rather than practice as physicians.

The ambition and focus of this particular group of students left an impression on me that had both positive and negative influence on my overall view of scientific research. My own research project on silicosis did not rank very highly among my medical colleagues. Blood chemistry, enzymology, and the big killer, cancer, were all deemed to be much more worthy subjects for research. This led to a certain arrogance from at least one of my colleagues, who engaged in overt questioning of my results, with a pedantic display of what I often regarded as supercilious comments. Also, it was at a time of the Arab/Israeli conflict (when is it not such a time?). There was one Iraqi who was in biochemistry with us. He held opposing views on the cause of the conflict and was naturally subject to much more acrimonious comment by the majority of Jewish students. One of the Jewish students, Paul, became a good friend of mine and we spent quite a lot of time together, sometimes visiting places around the island of Montreal. He had a great sense of humor and was the kind of person to whom I seem to gravitate more than others. His presence in the laboratory was a redeeming factor, which counterbalanced some of my relationships with others.

Student life in Montreal in the mid-fifties gradually became more pleasant. I had a small apartment near the university, which I shared with another student, who was also in biochemistry. We got along just well enough to live together but had different social tastes. I spent some time with Paul and others in coffee houses and, occasionally, at the Café St. André (known as Brother André's Shrine), which was a favorite drinking spot for university students. Very often, I went out on the weekends to visit my friends who had stayed on at Macdonald College. Now, I envied the beautiful campus setting with its willow trees and spacious lawns, compared with my dingy apartment in the city. I usually stayed with Herb McCrae and his wife. It was very generous of them to put up with me because they only had a two-room apartment, and I had to sleep on the chesterfield. Another friend, Don Layne, was now warden of the men's residence, and together with wives and various girlfriends, we usually had a pretty good weekend. Some weekends I stayed in town to work on my research, and once or twice I went down to Glen Sutton, Quebec, where the Johnsons had a country place.

It was during the latter part of my graduate studies that I met Anne

Hodge, who was a graduate of McGill and who came from a family in St. Lambert, Quebec. She invited me to her parents' house on several occasions, and I enjoyed the relaxed atmosphere, as well as her mother's good cooking. Anne's family had been Canadians for several generations and her grandfather on her mother's side had been chief counsel for the CN railway at one time. Anne was a pretty blonde; a caring person with an unselfish disposition. Our relationship grew into an engagement. The summer I obtained my PhD, I arranged for Anne to come over and meet my mother in England. (Actually, Anne did the "arranging" because I was penniless and relied very much on her savings for the next six months.) Anne and my mother seemed to get along well. I was relieved because I felt that my mother could be quite selective when it came to her son's future wife. Anne and I were married in the fall of 1958, and left Montreal almost immediately. We traveled by train across Canada to the west coast.

The reason why we were headed for the west coast lay in certain events some six months earlier, when I was looking for a job, following my doctorate degree. Most of my colleagues in biochemistry sought out further studies at prestigious institutions, such as the Carlsberg Foundation in Copenhagen, the Sloan/Kettering Institute in New York, or the University of Chicago, which boasted having the largest number of Nobel Prize winners in the world. Because of the highly competitive nature of biochemical research, however, I had become a little disenchanted with the field, at least, as one in which to make my career. I thought back on the days when I was involved in food science, and an opening for a chemist became available at the Department of Food and Drugs, in Ottawa. This still was not quite what I had in mind. I had heard that some scientists made a career in marine science, and that, possibly, I might be eligible to do the same. This would bring me full circle from my boyhood love of the seaside, to a career where I could actually be paid to carry out a hobby.

In the latter part of 1957, I wrote to a couple of famous institutions, including the Scripps Institution of Oceanography, but the reply was invariably the same, that there were no openings—the "keep-your-letter-on-file" type of response. Then, I saw an advertisement in *Nature*, asking for someone to conduct biochemical studies on phytoplankton with scientists at the Pacific Oceanographic Group, in Nanaimo, BC. This was it! A gust of wind out of the blue had brought me what I wanted. I pulled out all the stops I could think of, with regards to my résumé and references, and sent the package off with bated breath.

In April of 1958, after being told that I was short listed for the job, I received the following short letter from Dr. Jack Tully, who was Oceanographer-in-charge of the Pacific Oceanographic Group, an offshoot of the Fisheries Research Board of Canada:

Dear Mr. Parsons:

I have today recommended your appointment from October 1, 1958, with the rank of Assistant Scientist, at a salary of $5,600 per annum. If your work is in keeping with your qualifications and endorsements I will recommend your promotion to Associate Scientist after one year of service.

Yours truly,

John P. Tully (Oceanographer-in-Charge)

It was the most exciting letter I had ever received. Dr. Tully was a physical oceanographer of some repute, and he had assembled a number of other scientists into a group, to study the military and fisheries use of oceanographic data. Jack Tully had a very loud voice, a wooden leg, and a raucous personality that was full of praise one day and damnation the next. I could have imagined him as the captain of a pirate ship; there was never a dull moment with Jack. The man I would actually be working for was Dr. John Strickland, and he sent me a much longer letter describing what the work entailed and giving me a warm welcome to the west. When I told my colleagues in biochemistry at McGill where I was heading, the general response was, "Where on earth is Nanaimo?" and "How could you give up medical biochemistry for phytoplankton?" (although most, including me, did not know exactly what phytoplankton were, anyway).

I was obviously destroying myself, scientifically speaking, and no one expected to hear from me again.

PART TWO

Seascapes

I do not know what I may appear to the world, but to myself I seem to have been only like a boy playing on the sea-shore, and diverting myself in now and then finding a smoother pebble, or a prettier shell than ordinary, whilst the great ocean of truth lay all undiscovered before me.

(Isaac Newton, 1642-1727)

V
Learning to Swim

Anne and I traveled across Canada by train. The first part of the journey was not too exciting geographically, but once we had left Calgary, and the mountains were all around us, the west became truly exhilarating. Huge slabs of rock reared up out of the green forest, to be topped with a covering of snow, like icing on a bun. One enormous mountain, called Mt. Eisenhower, was a particularly fitting tribute to a man I had only previously encountered in England, where, as Commander in Chief of Allied Forces, he had reviewed troops on our school grounds before the invasion of continental Europe. The train continued to snake its way through tunnels as we ascended and descended the passes. Then came the trip down the Fraser Canyon; it all made for a wonderful honeymoon for first time adventurers. We crossed the Strait of Georgia on a dilapidated CP ferry and arrived in Nanaimo, to be greeted by a tall, lumbering, red-haired gentleman, Dr. John Strickland.

From the moment we met, until we parted four years later, John was everything a boss should be—kind, considerate and dynamic. He listened to ideas, developed them further, or told me why he did not like my opinion. He was my mentor for the first four years of my career in oceanography, and I owe him a great debt of gratitude for getting me started in the right direction. His own background was that of a chemist, from London University. He later went on to start a new department at the Scripps Institution of Oceanography, in La Jolla, California. He died, prematurely, in 1970, after establishing a strong research record based on quantitative relationships, linking the plankton community to the chemistry of the sea.

The work that Dr. Strickland had been hired to undertake was to start a new group within the Fisheries Research Board of Canada. Both the Director of the Nanaimo station, Dr. Needler, and the Chief Oceanographer, Dr. Tully, felt there was a need to obtain some understanding of the environment in which fish lived and flourished, to the extent of forming very large populations. Most of the research conducted up to the late 1950s had been centered on the populations of fish, and not on their environment. It was a brave new beginning to start to research the ecosystem of the fish. Dr. Strickland formed a small group of scientists within oceanography, and I was privileged to be a part of this research team. This mission, to understand the physics, chemistry, and biology of the ecosystem of fishes, was to occupy my whole scientific career with the government, UNESCO, and finally, with the university.

The name of the town, Nanaimo, where we now took up residence, actually means "the meeting place" in native tongues. This is a great source of amusement to Japanese speaking persons. In Japanese it means "Seven Potatoes" (nan=seven, imo=potato). It was formerly a coal mining town, but all the mines had closed by the time we arrived, in 1958. It had become a distribution center for the northern part of Vancouver Island, due to the regular ferry service to Vancouver and the fact that Nanaimo was a sea-going port. Otherwise, not a lot happened in Nanaimo. There was some sport fishing, occasional concerts were held at the high school, and there was a shopping centre. This lack of luster was recognized by the town's mayor, Mr. Frank Ney. In an effort to attract more tourists, he started a bathtub race across the Strait of Georgia. It was a strange coincidence to encounter a film of this race years later, featured on Chinese television, on a night that we landed in Beijing.

When I first arrived in Nanaimo, I took up cricket again. Our cricket club tried to interest Sir Dudley Stamp, a famed geographer at the London School of Economics, into becoming a patron of the club. He owned an island right in front of the Biological Station where I was working and my family happened to have lived practically next door to him in England. He was elected Honorary President of our club and told us he was "rather touched" by the honour, which must have been a very minor event for one who had received so many significant national and international accolades

I played with the club for a couple of years but later left, to take up tennis more seriously.

My first research work with John Strickland was to develop a manual of seawater analysis. He felt there were no standard methods for analyzing seawater, and that in order to obtain comparable results among oceanographers throughout the world, a basic text on analytical methods was required. The text, entitled *A Practical Handbook of Seawater Analysis*, was produced as a joint effort, and for at least three decades following its publication, it was considered the standard text for seawater analysis in the world. I have come to think that John's emphasis on methodology was eventually one of several key points in my later being awarded The Japan Prize. After John died, I used the same text as the basis for a teaching volume, *A Manual of Chemical and Biological Methods for Seawater Analysis,* by Parsons, Maita, and Lalli, and this too was very successful. It was translated into Chinese in the 1980s by the People's Republic of China (PRC). However, when later we tried to use the book in Taiwan for teaching, it turned out that the Chinese written in the PRC had changed so much from the original Chinese, as used in Taiwan, that the Taiwanese students were unable to read the text.

As an oceanographer, I was required to go to sea, and this was a new experience for me. Oceanographic research vessels are not large, usually less than 1,000 tons, although a few of the larger vessels are several thousand tons. The first vessel I went on was called the Canadian Naval Auxiliary Vessel (CNAV) *Whitethroat*. It had been converted from a cablelaying vessel into a research vessel. My first cruise was into Queen Charlotte Sound, a body of water similar to the Bay of Biscay, in being notorious for its rough seas. We were not to be disappointed. The ship seemed constantly to stand on end and come crashing down again with tremendous shudders to the superstructure. Scotty, the chief engineer, assured me several times that we were unlikely to sink. I hadn't thought about sinking until he kept mentioning it, and then I wished I was back with the black flies in Quebec. I was violently seasick and retched constantly into those stormy coastal waters of the Pacific northwest, as did my fellow scientists. Nonetheless, we managed to collect some samples and from these samples we determined the salinity and temperature, which told us where the water had come from. We collected plankton to measure the production of the water, and we preserved species for later examination on shore. The person collecting the samples had to stand on a platform outside of the hull of the vessel and place metal sampling bottles on a hydrographic wire that descended into the depths. The platform on which he stood was known as "the chains," partly because it was surrounded by chains for safety and partly because the occupant had to be "chained" in with leather straps, in order to prevent his being washed away. To be "working in the chains" could be quite an adventure. In spite of the stormy conditions, I was enraptured by my first oceanographic cruise.

When the data were put together, I was to see for the first time that the water column had some structure, with temperature and salinity differences between the top and the bottom, and superimposed on this were differences in the distribution of the nutrients and plankton. An echo sounder on board showed us how the zooplankton migrated to the surface layers during the night and then sank again away from the dawn light. At night, many luminous creatures raced through the waters.

We were on a twenty-four-hour day with watches; four hours on and four hours off. Collecting oceanographic samples is a very expensive business and we could not afford to let the ship be idle for half a day, but it was physically very tiring to endure both the continual motion of the boat and the lack of a good night's sleep. When we came ashore, the chief oceanographer on the cruise, Mr. Dick Helinveaux, had to write a report on the performance of his seasick crew. He was very kind and, notwithstanding my physical state, he reported to Dr. Tully that I would make a good oceanographer.

Not long after this experience, I embarked on a second expedition, which John organized. He managed to get us on to the Department of Transport weathership, the *St. Catherine,* which occupied a position for seven weeks at 50°N, 145°W in the Gulf of Alaska. Very few studies had been made in the open North Pacific and John was particularly interested in how these waters compared with the North Atlantic at the same latitude. Why, for example, was the Pacific always called the "Blue Pacific" and the Atlantic was never called the "Blue Atlantic"? It seems a trivial question, but one which was certainly answered by our studies. The amount of plankton in the Pacific waters at our latitudes is much less than is found in the North Atlantic, the latter having been well studied for several decades. The lack of plankton in the North Pacific may be related to several mechanisms, but it has the effect of allowing blue light to be reflected back to the eye, while in the North Atlantic, the small planktonic particles scatter light, which results in a greenish tinge to the water. The purpose of our study was to find out why, at the same latitude in two different oceans, there were two different production mechanisms. When we published our ideas on this, a renowned oceanographer from Europe commented, "There must be something wrong with the Pacific!"

What he really meant was that the Atlantic had been studied earlier, and a theory laid down on seasonal changes in the Atlantic Ocean, which were assumed to be the same in the Pacific. They were not, and what was "wrong" (different) is still giving rise to much scientific debate even now, forty years later.

My experience on the weathership was much more tranquil than my first expedition into the Pacific. The *St. Catherine* was an ex-navy frigate, and she rode well in the ocean swells. Frigates were designed by William Reed of the United Kingdom, following his success in designing corvettes. He designed the latter after the whale-catching boats of the Antarctic whaling era, and they were made especially seaworthy for those tempestuous waters. They compensated for the lateral force of waves by rolling sideways and slithering through the heaviest of seas. The frigates were much more powerful versions of corvettes; they had twin screws and could stay at sea for much longer periods. The *St Catherine* was built in 1943 and saw action during the war in protecting North Atlantic convoys, which included assisting in the sinking of at least one U-boat, the U-744.

Every day, as we maintained our weather watch and conducted oceanographic experiments, we were circled by black-footed albatross and, occasionally, by the white Laysan albatross. Storm petrels and phalaropes flitted around on the sea surface, eating plankton without ever landing on the water. An occasional gray whale passed us on its way to the Gulf of Anadyr for its summer feeding. Above all, I remember the intensity of the blue waters; to come back to the coast after seven weeks and see something green, was an extraordinary visual relief.

Because we were at sea for such a long time (longer than it took Columbus to reach North America), we experienced some of the social upsets that accompany an isolated group of men under any similar circumstances. The jovial attitudes we all had at the beginning of the expedition tended to sour with time. It was not long before the small group of oceanographers was regarded by some of the crew as a bit of a nuisance to their daily operations. However, apart from some words and looks, nothing serious happened, but we all relished the moment when the relief weathership appeared and we could head for home.

In the early 1960s, John had the idea of encapsulating a large volume of seawater, to follow changes in the plankton and nutrients under near natural conditions. It is hard to follow such events in the open ocean because currents are continually carrying away the plankton as they are produced. By making a transparent plastic bubble, containing about 100 tons of seawater, we were able to follow the dynamics of a phytoplankton bloom and its subsequent consumption by zooplankton. This technique represented a first in ocean ecology. These large plastic bags became known as mesocosms. About ten years later, I was to continue these experiments with much greater volumes of seawater

In addition to these experiments, I embarked on an analysis of different phytoplankton species, in order to find out the nutritional value of various kinds of plankton blooms. The paper we wrote on the chemical composition of phytoplankton became a citation classic, which meant that it was frequently cited by other authors, as determined by an independent abstracting agency. This was a significant period. The discoveries of new data and methods were providing a firm foundation for my future work.

My work with John Strickland was not without its difficulties. About halfway through our cooperation we moved into a new building with much better facilities. Someone did not like us there because one day, we came to work to find an experiment that we had set up was smashed—broken glass and cut wires—no accident, but sabotage by another scientist. There was more than one occasion on which we encountered this kind of event, and although the RCMP were called in to investigate, there was little they could do. Since the locked rooms were only accessible to ourselves and one or two very senior scientists in administration, it was fairly obvious who was involved in these shenanigans, though nothing could be proven. Worse things have happened in some other laboratories, and such events occasionally get into the press. I believe that the reason for this is that scientific discovery is very competitive to some persons, and to see new ideas overtaking one can arouse the baser instinct to find a way of putting down the competition.

Every summer we had an intense research program, which sometimes involved going to sea or working on John's plastic bubble. We always took on a group of undergraduate students as assistants, to help with our summer research. On one occasion, this resulted in a near disaster, when we sent two of the students down to refill the navigation lights on our experiment located in the sea. They had filled the kerosene lamps with the appropriate fluid and then started to check each one by lighting the wick. Unfortunately, they had spilled a small amount of kerosene on the dock and this ignited; it could easily have been extinguished, were it not for the very peculiar habit of fire marshals, who insist on painting both fire extinguishers and gas tanks the same color, red. One of the students mistook a red canister of gasoline as being part of the fire-fighting equipment and started to dump the contents on the otherwise innocuous flames—they both just had time to dive into the sea before the whole wharf erupted in flames. This is the sort of mistake one does not want to make when out at sea, with little help nearby. Fortunately, the local fire department managed to get everything under control and, apart from a scorched dock and the loss of the director's sailing dingy, no further damage was done.

After four years of fascinating studies and other experiences with John Strickland and his colleagues at the Pacific Oceanographic Group, I decided to try my hand at international science and applied for a vacant position with the United Nations Educational, Scientific and Cultural Organization (UNESCO), in Paris. There was an Office of Oceanography within this organization, and appointees were also members of the Secretariat of UNESCO. Once I was offered the job, it was not difficult for me to exchange the culture of Nanaimo for a life in Paris. The world is seven-tenths ocean, and all I had seen in four years was a little corner of the North Pacific. It seemed to me that an international office of oceanography should satisfy my appetite for knowledge on a global scale.

Meanwhile, on the domestic scene, we had two lovely daughters to take with us to Paris. Stephanie was born in 1959, and Allison in 1961. Allison's birth was particularly eventful. She was born at home, and I was the only "doctor" presiding. My wife had slept through labor. We hurried to the bathroom, where Allison slid into my hands, bellowing loudly. I felt proud to be the first to hold the tiny baby as she emerged. My wife remained very calm, and when the physician arrived, he took mother and child off to the hospital. They were not allowed to be in the maternity ward with other mothers and children because the birth had not been conducted under the sanitized conditions of the hospital. I was a little hurt that they were isolated for a few days before being allowed home again.

    The children were one, and two-and-a-half years old, when I left for Paris. I went a month ahead of the family, in order to find an apartment. I rented a three bedroom unit on the Rue de Chaillot, not far from the Arc de Triomphe and only ten minutes from UNESCO headquarters. I arrived in Paris at the time of the Algerian troubles, when General de Gaulle had decided to give up trying to govern Algeria, but it was expected that some French generals might try to pull off a *coup* at any moment. On my second night in Paris, I was awoken by explosions and flashing lights. It seemed that I had arrived in the middle of a revolution, but there was little panic in the street below. On further inspection, I found that the Eiffel Tower, which was not far from my hotel, was alight with fireworks. All of this was in aid of the premier night of a film, *Le Jour le Plus Long*, which had started that evening at the Trocadero. It turned out to be a pretty long day for me, too! Fortunately, this was the closest I ever came to being caught up in a revolution, although some travelers are not so lucky.

VI
Sailing in International Waters

I happened to arrive in Paris when all of the professional staff from the Office of Oceanography at UNESCO were away on a mission. It was a lonely week, before they dribbled back into town. My French was practically non-existent, since I had learnt German in school, so that even staying at a small hotel and ordering meals was an exasperating experience. Fortunately, in the office, my boss was American, his second-in-charge was Russian, there was one Japanese, and myself, which dictated that the common language had to be English.

UNESCO headquarters were located in a modern building, opposite the Ecole Militaire. I checked in with one of the multilingual secretaries sitting at a long, raised desk by the entrance. I was told the floor and number of my office and was greeted there by an English secretary, thin-faced and domineering. She had me slotted in as the most junior member of the oceanography professionals, and one who was obviously not to be taken seriously. From an administrative point of view, she was efficient to the point of often taking control of discussions with my American boss. She certainly had far more input on policy than I did in the first year of my tenure. This was really not surprising because the daily operations of UNESCO were more an in-house political power struggle with the administration, for attention of the Director General, than it was a professional approach to the science of oceanography. Our senior secretary was street-wise to the internal operations of UNESCO.

Visitors from various countries came and went, seeking UNESCO's approval of some plan. This approval was far more important to obtain than any part of our small budget. The stamp of UNESCO approval authenticated the desires of our visitors, which, in turn, opened much bigger financial doors to their ventures than we were able to offer.

My work with UNESCO had virtually no research component, and might be described as almost entirely administrative. My projects included the setting up of a plankton laboratory in India, arranging for a training course for young oceanographers in West Africa, attending meetings of other fisheries and oceanographic organizations (particularly the Food and Agriculture Organization—FAO of the UN in Rome), and taking part in some work on standardizing oceanographic methods. The last undertaking was the only one which involved some science.

In addition to these activities, I played a minor role in the organization of oceanographic expeditions, notably in the Indian Ocean. This was a very important part of oceanography at UNESCO because it brought together the Russians and the Americans during a period in history when the Cold War was at its height. In fact, we were the liaison office for the Cooperative Investigations of the Tropical Atlantic—a program involving both Russian and American vessels during the Cuban crisis, when Mr. Kruschev wanted to unload missiles just off-shore of the United States. During this period, the oceanographers of the two nations continued their investigation of the tropical Atlantic, even as Russian transports and American warships jockeyed for position in about the same area. This was an example of how, on more than one occasion, the United States and Russia agreed to carry out oceanographic cooperation, while still eyeing each other with grave suspicion.

Sometimes, there were misunderstandings between the two nations, even within UNESCO. An amusing incident occurred during the first Space Conference, which was held at UNESCO, and which I attended as a member of the Secretariat of UNESCO. Because the Russians were first into space, their delegate opened the conference with a speech about the glories of Soviet rocketry. The Americans were allowed to ask the first question.

Sailing in International Waters

"Has the Soviet Union solved the problem of defecation in space?" asked their delegate.

UNESCO translators were pretty good, but this word "defecation" stumped them. The Russian stood waiting for a translation of the question. People who had microphones started to switch them on. To be helpful they shouted "shit," "crap," and other graphic words. The translator interpreted this as the American delegate's verdict on the speech! Hearing this, the Russian got very red in the face and stormed off the podium. It took some time for cooler heads to prevail and relay the correct translation.

On another occasion, at the time of UNESCO's first meeting of the Intergovernmental Oceanographic Commission (IOC), there was supposed to be an exchange of introductory speeches from various delegates. Dr. Warren Wooster, who was the secretary of the IOC, was sitting next to Dr. Konstantine Federov, his second-in-command, when the Soviet delegate, Admiral Rychkov, made his welcoming address. The Admiral's speech was a bitter attack on the West, in which he used clichés such as the "Imperialist Powers," "Warmongers," and other derogatory comments. Dr. Wooster was most alarmed at such an opening presentation and quietly asked Dr. Federov what was going on.

"It is OK," said Konstantine, "this is normal Soviet speech for an international conference – after it is over, we will have a good meeting!" – and he was right.

During my UNESCO years, I benefited from making contacts with people whom, otherwise, I might never have met. Among these were Dr. Cushing and Dr. Steele of the United Kingdom, whose works on plankton and fisheries I greatly admired. Dr. Cushing was about ten years my senior. He was a biologist who wrote extensively about plankton. Dr. Steele was a mathematician, one of a very few in biological oceanography at the time. He produced some of the first mathematical models of marine ecosystems. Professor Hempel of Germany had excellent ideas on fisheries, although he spoke to everyone as if he were the Kaiser. Professor Uda, a physical oceanographer from Japan, introduced me to ideas of fisheries oceanography, that I was later able to develop further. Mr. Sidney Holt, of FAO in Rome, was the chief architect of a theory on fisheries management that I came to despise, and there were many others with whom I could agree or disagree, with respect to their science. This was of great benefit in widening my view of ocean science.

Another special aspect of life at UNESCO was the opportunity to travel. We were periodically sent out on missions, which were trips away from Paris, in which we engaged in some activity for the organization. I traveled twice to India and once to West Africa on missions, and attended meetings in most of the western capitals of Europe.

India was the most fascinating country, full of so many different people and wonderful ancient architecture, but run by a bureaucracy that defied imagination. Most international airlines arrived in Bombay airport in the middle of the night, yet stepping off the airliner was like walking into a Turkish bath. However light my clothing was when leaving Paris, it was not light enough for the stifling heat and humidity of Bombay. Immigration officials sat at long rows of desks, and my passport was inspected several times. There was a dank smell of sewage, partially masked by the odor of an insecticide. Airport lighting intensity was little more than bright moonlight. Porters eagerly tried to assist me with my one bag and were astonished, bewildered, and emotionally hurt when I declined their attention. More teams of customs officers inspected my bag and, finally, I was allowed the freedom of a personal taxi to the Taj Mahal Hotel.

The hotel was on the waterfront, and I had a magnificent view of the bay. It had most of the modern amenities but later, when I fancied a beer before dinner, I ran into what was then an era of prohibition on the subcontinent. However, I was told that if I could obtain a special passport from the police, indicating that I was a guest of the Indian government, I would be allowed into the hotel bar, which was supposedly only for foreigners. The police desk was conveniently located outside the bar and, with my special permit, I entered a cavernous room whose sole occupants seemed to be Indian nationals. When I asked one of the patrons about this, he said, "Oh very simple, Sahib. Drinking permits can be issued by doctors for medical reasons."

It seemed to me that everyone looked perfectly healthy, but this was probably because all their ailments subsided once they began to tipple.

In the 1960s, Indian National Airline bookings were often shambolic. Even with confirmed bookings, it seemed that we usually had to have a second reason for boarding, which had to be better than someone else's. The one that I used was that I represented considerable UN monetary aid for India and that if I missed my meeting, India would miss its aid. This usually impressed the airline desk attendant.

Sailing in International Waters

I had to travel from Bombay to Cochin, in the south, where my mission was to aid in setting up a Plankton Sorting Centre for the International Indian Ocean Expedition. We took off from Bombay on the regular four-hour, daily flight in a two-engined, piston-driven plane. We made a stop halfway down the coast. There appeared to be nothing to stop for, just a runway, and a man holding a small fire extinguisher, standing at the end of the ramp. It seemed that since there were no toilets on board, we had stopped to allow everyone to relieve themselves, which we did, men on the right and ladies on the left, and then we re-boarded for the rest of the flight. The inconvenience of travel was made up for by the wonderful Indian curried food, which we enjoyed at our destination in Cochin.

The scientists at the Center were generally helpful in explaining their work and needs, but there was a hierarchy of scientific opinion which sometimes prevented us from finding out what the problems were at the grass roots. There was also a curious importance attached to the size of the desk assigned to each person. Junior scientists had desks they could barely get behind, while senior scientists had much larger desks, but it was not until we met the Deputy Minister of Science, in New Delhi, that we encountered the ultimate desk. It appeared to be about half a tree, sliced sideways, superbly polished, and long enough for our whole delegation of fourteen persons to sit at, in one, long row. The minister looked rather lonely on the other side, but in the Indian hierarchy, his position signified great importance.

Making sure that UNESCO aid reached the desired location was sometimes difficult. We had ordered a lot of equipment for the Plankton Sorting Centre in Cochin, but some of it had been diverted to other laboratories. At the time I visited Cochin, there was an Indo/Norwegian fisheries project in that town that was much more efficiently run than our UNESCO project. The reason, I believe, was that the Norwegians stayed on site in considerable numbers, and assured the arrival and distribution of all aid. UNESCO trusted the Indian bureaucracy to do the same thing, and that was a mistake. It convinced me that bilateral aid was more efficient than multinational aid, because in the latter, there was no focus on responsibility.

I was also sent on a mission in West Africa, to find a suitable location for a regional training course in oceanography. I visited Cameroon, Nigeria, Ghana, the Ivory Coast, Guinea, Senegal, and Sierra Leone. Sometimes, when I woke up in the morning, I was not sure which country I was in. Also, it was the only area in my travels where I found French to be essential, for in four of the countries the alternative language was not English but some African dialect. By then I spoke a little French and managed to interview various scientists and government officials.

I found that the most bothersome aspect of missions to the tropics was how to cope with the multitudes of insects that seemed to be such a part of every day life. In India, it was the flies that stuck to your face while you were eating, and in West Africa it was the cockroaches, and swarms of termites that could march into the marbled halls of even the best hotels. I remember being met, at one airport, by a very officious French official who was explaining in great detail the program he had set up for me, when I saw an enormous cockroach scurrying across the floor directly behind him and in a line for his feet. I was fumbling for the French words for "take care" and "cockroach," both of which I knew, but I did not know how to say that it was about to go up inside his trouser leg. A couple of seconds later, the roach disappeared, and my colleague's detailed instructions were terminated with a "Mon Dieu!" and a distinct effort to protect his crotch. The harmless creature was finally dispatched with a slap on the thigh which, in turn, produced a noticeable stain of insect body fluid on the official's trousers.

In Accra, Ghana, I stayed at the Black Star hotel, not far from a sports stadium where Dr. Nkrumah had been shooting up his opposition. I don't think there was ever such a promising African leader who became so politically corrupt as Kwame Nkrumah. He inherited one of the most prosperous countries in Africa, but his Marxist economic theories were a disaster for the country. In February of 1966, he was finally ousted from office in a military *coup d'état*, known as Operation Cold Chop, and Colonel Kotoka started to lead the country toward recovery.

I learned, in Accra, that one should not walk about alone in some African cities, even in broad daylight. Just after arriving, I wandered out of the hotel and walked down towards the market to take in some of the local scene. Several Europeans stopped their cars near me as I walked along the street and asked me if my car had broken down, and did I need a lift? I should have been street-wise enough to have known that I was somehow out of place. On reaching the market, I was suddenly engulfed by an angry trio of rather ancient Ghanaians who told me to go home, no white boss wanted now! I hurriedly retreated back to my hotel.

It was on my mission to Accra that I came across the ultimate inefficiency of the UNESCO system. I checked into the hotel in Accra, where there was a UNESCO regional office that was run by an expatriate Englishman of the rather snobby type. He was tubby, and his white shirt strained every button over his enlarged abdomen. He lounged in a comfortable chair behind his desk, smoking a cigarette, while languidly giving orders to various clerks and making silly little aside comments to another European employee, seated on an upright chair nearby. I came into his office, already tired by the heat, noise, and confusion of a large African city. After a carefully contrived period of keeping me standing in front of his desk, he said, "Good heavens, Dr. Parsons, why on earth didn't you tell us you were coming? Can't expect us to know everything you know."

"I filed my mission through the Bureau of Relations with Member States at UNESCO, Paris, and it is their job to inform you," I replied, wanting to sit down and be offered a cool drink.

Actually, it was the role of the regional office to facilitate the visit of persons from headquarters. The question seemed to me misdirected: it should have been "What can we do for you?"

Things did not get any better when I asked for a car to go and visit the Secretary of Science for Ghana. I had seen three large black limousines parked under a shady tree outside the headquarters, and imagined that I could get the use of one with a chauffeur for a couple of hours. Later, I found out that they were used exclusively for transporting the office director and his staff home, for lunch and dinner.

"Well, actually we do not have any cars available, old boy. We might, however, find out if the mail truck has finished its run for the day," said the director.

With a flurry of pseudo-businesslike phone calls he deduced that the truck was indeed available, and I boarded the most ramshackle vehicle that I had seen for some time. We bounced over the roads to the Ministry of Science, a large, red building, surrounded by trees laden with white blossoms. I shook the dust off my suit and was in time to see the Secretary of Science. He waved me into his office, where a huge ceiling fan gave some relief from the stifling atmosphere. I had his name written down as James Joseph but he had Africanized it to Yanni Yoséé. I explained to Mr. Yoséé how we were thinking about holding a training course in one of the West African countries, and he sounded most eager that it should be Ghana (of course!). After our meeting, the mail truck had disappeared, but I was fortunate to get a ride back to my hotel in a vehicle provided by the Secretary of Science.

The next day, I went to the German and American Embassies, to find out if they knew anything about the state of marine science in Ghana. These two embassies always knew what was going on in any country, and I never went to any others, including our own Canadian embassy, because I had found that they did not know much about their host country. At one of the other embassies I was told that corruption was rife. Some UNESCO tape recorders for schools, provided by the "E" in UNESCO, had all been distributed among cabinet ministers, as gifts from the Minister of Education. So, this was UNESCO in Africa; a regional office run by a pompous 'Brit' who "couldn't care less" about international aid, neither helping visitors, nor checking out the distribution of material aid when it arrived. It was only a few years after I left UNESCO that the United States decided to withdraw from the organization entirely. A whole series of Directors General of UNESCO had failed in their administrative responsibilities to uncover the utter waste in the implementation of programs.

In Abidjan, on the Ivory Coast, I had chosen to stay at an hotel with the name, l'Hotel sur la Plage. I thought that it would be a good choice, where I could listen to the waves on the beach at night and go for a morning stroll along the shoreline. Little did I know that la Plage was where, it seemed, half the population of the city came to defecate, early each morning. I changed my hotel after one night. In spite of this, the city and people of Abidjan were a pleasant surprise, after the industrial confusion of Lagos and Accra. I felt that it was at least one place that could host an international training course for marine scientists.

From Abidjan, I flew to Dakar, which is sometimes called the "Paris of Africa." A largely Moslem city of about half a million inhabitants, it was refreshingly free of fundamentalists. The women of the city were superbly costumed in flowing, colorful robes and turbaned headdresses. They ran an assortment of gift shops, cafés, and bars, spread out among the elegant houses of the town. Baobab trees and bougainvilleas added colorful vegetation to the city streets. About 4km offshore is the Ile de Goree. The island has been fortified by several generations of colonial rulers and was the principal center for the export of slaves from the 15th to the 19th century. I visited the Maison des Esclaves, where slaves were confined before shipment to the New World. It is a politically sensitive museum, and it was an emotional experience to be brought face to face with this part of history.

In Dakar, Senegal, I stayed at a wonderful hotel which had a bar, the roof of which seemed to be a series of very large umbrellas turned upside down to catch rain, rather than to dispose of it. Each umbrella had a hole in the centre, and it was not until about 6 p.m. that I discovered what this was for. Practically every evening, on the coast of West Africa, there would be a thunderstorm that lasted for about an hour. During this time, the rain came down in buckets and was collected by the rather unique roof of the bar. The hole in the centre of each umbrella then became a miniature waterfall that emptied into a small pool which overflowed, sending a little stream flowing out of the bar. It was a very pleasant sensation to watch and hear the water cascading down, while quaffing a beer senégalaise and listening to the deep roar of thunder in the background.

My contact in Senegal was supposed to be Professor Monod, who was head of a sophisticated research centre, called L'Institut fondamental d'Afrique Noir. He was unable to see me on my first day in Dakar, so I took a ferry over to the Ile de Goree. My main purpose of visiting Goree was to see a small marine research laboratory as a possible facility for a training course. I was well entertained over lunch by its Director, who was, however, quite alarmed that I had not yet been able to contact Professor Monod. The next day, when I was ushered into a large and comfortable office at l'Insitut, Professor Monod started our conversation in French, by castigating me for visiting his subordinate on the Ile de Goree without his permission. It had not occurred to me that the professor ruled his little kingdom of scientists with such rigid autocracy. We never managed to hit it off after my *faux pas*, and while I greatly enjoyed the ambiance and people of Dakar, I did not feel it was a place for an international training program.

I visited the same city twenty years later, on another mission for UNESCO concerned with coastal research. By then, most of the French had left, including a garrison of French troops. The city had become very dangerous for white people in the 1980s. It was not only the danger on the street if one was alone, but at the airport, when leaving for Paris, I was taken off to a small room by a heavily armed policeman who demanded that I empty my wallet and pockets of all my money.

"You are not allowed to take money out of Senegal," he said.

I was, most fortunately, rescued from this one-on-one encounter by an Egyptian colleague with whom I was traveling. He just burst into the little room and told the policeman that I was a very important person and all hell would break loose if he detained me. I owed my Egyptian friend, Professor

Joseph Halim, at least a night out on the town, in Paris, when we returned. His streetwise ability to turn the tables on a heavily armed thug was quite remarkable.

During my tenure at UNESCO, I had two bosses. The first was an American, Dr. Warren Wooster, who was a very humorous and likable person, and a strong supporter of the concepts of fisheries oceanography. He and his wife helped our family to get settled in Paris, and he has continued as a good friend over the years. Unfortunately, although he was both politically and scientifically well motivated, he never gave me anything interesting to do. Things changed when he left and a Russian, Dr. Konstantine Federov, took over the office. He was a rather short person, very dynamic, and full of carefully thought out ideas. I was given much more work, and the West African trip was one of the projects that he assigned to me. When I wrote my report, I suggested that either the Ivory Coast or Sierra Leone was the best place to hold the training course.

"Now about this report," said Konstantine, "you must understand Tim (although he always pronounced my name 'Teem'), it is our duty to help the poorest African nations. Guinea is where I would like to see the course held."

"Well, as I pointed out in my report, Konstantine, Guinea has recently declared complete independence from France and the country is in chaos. As an example," I explained, "a research vessel which was in fine shape when the French left, was sitting on the bottom of Conakry harbor, still tied to the wharf, but six feet under at high tide, for total lack of any maintenance. It would be a waste of money to hold a training course in that country."

"Ah, but now you have stated the very reason why we should hold the training course there. It will help to stabilize the country, now that the imperialist power has departed."

I could understand Konstantine's point, in an idealist sense, but having seen the living conditions in Conakry, I could not agree with him from a practical point of view. We continued to have a noisy but pleasant argument about this issue. I was never sure how it was finally resolved because at the usual pace of happenings in UNESCO, no decision was made about the course before I left, a year later.

Back in Paris, my two daughters were growing up. They alarmed us greatly, one day, by consuming a whole box of children's aspirin between them; they had to have their stomachs pumped at the American hospital. We spent part of the summer in the country that year because a very lovely French/Jewish couple lent us their house at Barbizon. The family, with our *au pair* assistant, Thea, from Austria, moved out for the summer, and I went up on the weekends.

The children did not learn much French during their two years in Paris, although my eldest daughter attended kindergarten at a French convent. One day, I asked her to tell me about her friends at school. She said that she had two special friends; one was called "C'est si bon" and the other was called "Bonjour." Trying to deal with the Catholic nuns and the French language at the same time was too much for her. She ended up being awarded the prize for the worst behaved student in the school. I am sure that *I* would have qualified for this earlier in life, but my school made no such awards. I thought it was a little unreasonable of the nuns to include this prize as part of the prize-giving ceremony at the school. However, I clapped loudly when Stephanie went on stage to receive it because I had missed the description of its nature in French. My wife, Anne, understood the description. She rightly took a very dim view of the procedure.

I was more annoyed with the nuns on another occasion, when I parked my car in a legal parking zone, but one which the convent sisters had marked with a "Private" sign, saying that it was for the convent only. Since we were visiting the convent, it seemed legitimate on both counts that I should use the parking space. When I came back, a very large piece of paper was glued to the windscreen glass on the driver's side, saying that I had no right to park in that particular spot. Since I could not drive home with the car in such a condition, I was forced to scrape the paper off my windscreen with a defrosting knife, a task that took me a good half hour. I developed a poor opinion of these sisters of charity.

Living in Paris, we found out how many friends we had. We acted as a magnet for anyone coming to Europe. People we had not seen or heard from for years dropped by, and we duly entertained them. Among these was the late Dr. Peter Larkin, whom I entertained with a *boeuf fondue* dinner and an after-dinner stroll through the picturesque environs of Montmartre. It is a subject for later discussion, but I was to have much more contact with this scientist, when I tried to integrate oceanography, fisheries and climate into a single entity, at the university.

Once we knew the streets of Paris, it was one of the easiest cities to get around via the Metro, and it was certainly one of the most entertaining. I preferred it greatly to either New York or London, although I did not know either as well. The wide, tree-lined boulevards and the very efficient traffic control made Paris a lovely city to drive around, to see the sights. The countryside of France is also a treasure, and we drove our little Caravelle from north to south. The real charm of European countries lies in their 'out of the way' places, and France has many of these, which are never mentioned in any guide. Among other European capitals, I would rate Oslo as perhaps my first choice, based on the people, the architecture, and the climate, which is much cooler than Paris. I have never been able to stand really hot places, like Singapore or Hong Kong, even though I was born only 5°N of the equator. I always feel that once the temperature goes much above 25°C, and the humidity climbs with it, my brain becomes addled, and beautiful thoughts, or indeed any thoughts, disappear. Cooling off becomes my obsession. However, in some places, such as Hong Kong, the extremely cold air-conditioning is just as unbearable as the heat. I have found that the best solution to tropical heat is a large fan.

I left UNESCO soon after completing my two year appointment. I saw no future in endless meetings, scientific missions which were ineffective, and a bureaucratic organization where many strove to keep their tax-free jobs, while living in a wonderful city. Appointments to UNESCO are supposed to be largely term positions, so as to allow for changes and people with new ideas. However, there was not enough turnover of staff to fulfill this ideal, and some programs just floundered, while employees were getting richer. When I was leaving UNESCO, I received an offer of an Assistant Professor position at Harvard. Perhaps, I thought, I could redeem my status with my former colleagues in the Biochemistry department at McGill. Harvard had some of the most prestigious biologists in the world. However, I saw myself having to teach very large first year classes in biology, which is often the chief role of a newly appointed assistant professor in any university.

I had a delightful visit from Dr. Jack Tully prior to my leaving UNESCO, and he advanced some more forceful arguments for my returning to the Pacific Northwest. Dr. Strickland had left for Scripps Institution of Oceanography, and Dr. Tully said that he was about to retire. There would be many opportunities for new research projects, and I favored his advice, agreeing to return to the Pacific Oceanographic Group, in Nanaimo.

My family and I traveled back to Canada by boat on the P&O liner, the *Orsova*. It was a wonderful month for all of us to enjoy the luxury liner and visit other places on the way home. The children were taken care of, most of the day, by crew trained in child care, so we were free to enjoy all the other amenities of the vessel. We stopped in Bermuda, Fort Lauderdale, Panama, Acapulco, and San Diego, where I was very pleased to meet my former boss, Dr. John Strickland. He was now in charge of a large ecosystem research group at Scripps, where the research facilities were much more in keeping with his capabilities.

VII
Captain of My Own Boat

The Pacific Oceanographic Group (POG) was part of the Biological Station in Nanaimo. Very soon after I returned, on the retirement of Dr. Tully, the POG was disbanded. The government agency managing the Biological Station was changed from the Fisheries Research Board to the Department of Fisheries. The change made employees part of the Canadian civil service for the first time. Hiring practices and other details were affected, which tended to make the research atmosphere more restrained and more directed by Ottawa. In spite of this, I was given a free hand to start a program in biological oceanography, and was allowed to form a small group of scientists interested in the same ideas. This feeling of scientific independence became a very important part of my future activities.

The ideas that I wanted my group to pursue were largely based on trying to integrate oceanography and fisheries, to give us an holistic understanding of how the biology of the ocean was driven by the physics and chemistry. Our small group had people experienced in plankton, fisheries, and chemistry. We lacked a physical oceanographer, but there were physicists at the Biological Station with whom we could consult.

The first task with any group, it seemed to me, was to have everyone understand that they were involved in a group exercise. I started this process by having all of us study a fjord called Saanich Inlet. It is a body of water near the Strait of Juan de Fuca and, consequently, it is flushed by ocean waters, but is sufficiently calm to allow us to perform experiments. Here, we set up our laboratory in an old naval barge, which was loaned to us for the purpose, and we started to study everything biological that we had time and space to measure. We followed the nutrients, the phytoplankton, the zooplankton, and the young salmon in the inlet, and we tried to develop a space/time continuum of how the different components of the ecosystem came together. We made some similar, but less intensive, studies in the much larger open waters of the Strait of Georgia.

The main purpose of these studies was to develop ideas on how the fish in the sea were controlled by their environment. The most popular theory on fisheries at the time said that the number of young fish entering the sea and maturing to adults was a function of the number of parent fish; it was known as the stock/recruitment curve, and it ignored environmental variables. This theory had been tested in management strategies for several decades, with only limited success. In the meantime, fisheries resources were declining, due partly, but not wholly, to fishing pressure. Since there was no understanding of why fish were so abundant in the sea, it did not surprise me that continued fishing would eventually lead to their demise. We had to understand the physical, chemical, and biological processes which made such fish as the herring, the tuna, the cod, and other fish superabundant, and why this abundance sometimes changed over time. To me, this had to be the adaptation of the fish to some properties of the environment. When these properties changed, one species of fish might decline, and other fish might become more abundant. There were few fisheries scientists who believed this argument, but among those who did were Dr. Taivo Laevastu in Seattle, Washington and, to some extent, Dr. Cushing in the United Kingdom.

Above all, it was the creed of Professor Uda in Japan, who had been preaching this philosophy (which became known as "fisheries oceanography") for several decades, that most impressed me. Professor Uda was a rotund figure with a smiling face, like a Buddha. He had a limited knowledge of English but could always make himself understood with various gyrations, as well as sketching his ideas on a blackboard. He held very strong opinions on how oceanography should develop. I once asked him what he felt of the idea of allowing some coastal zones to have different pollution standards to others, which were less directly in the pathway of urban effluents.

"Very silly idea," said the professor. "Ocean go round and round; pollutants go round and round, too!"

Identifying my research with some of these well-established oceanographers, as well as with some younger scientists, was an important step in my career.

In our work, we really needed a physical oceanographer, but most of the physical oceanographers on the west coast in the 1960s were reluctant to give up their studies of physics to look at fisheries problems. They found it too distracting to take on the much more difficult science of the biology of the sea. It was not until a decade later, when we had a new generation of physical oceanographers, that they started to devote themselves to fisheries problems with almost instant scientific success.

Our work on the chemistry and plankton of Saanich Inlet and the Strait of Georgia gave us a good indication that these waters were healthy, and functioning as a robust marine ecosystem, despite some pockets of pollution around pulp mills and urban areas. The residence time (i.e. the time required to replace the water) in this coastal area was generally less than two years, which did not allow for the build-up of toxic pollutants. With our findings, however, Minister of the Environment Jack Davis decided to go on a publicity blitz about coastal pollution. He invited Mr. Jacques Cousteau to fly over the Strait and express his opinion on pollution. From at least 5,000ft in the air, Mr. Cousteau reported that everything would be dead in twenty years, if we did not clean up our pollution in the Strait of Georgia! These were newspaper headlines, and just what Mr. Davis wanted to hear. Inadvisably, since Mr. Davis was now my boss in Ottawa, I wrote a rebuttal to the newspaper recounting our belief that the Strait was in pretty good shape. My letter was duly published and I, in turn, received a verbal reprimand from the Assistant Deputy Minister. Twenty years after this prognosis, it has been largely forgotten, but it is an example of how politics can put a different spin on reality. Scientists are seldom elected to our parliament. Is it just because they are a dull lot, or are they unable to compete in a world that is mostly dominated by lawyers and labor leaders, who have a natural ability to communicate with crowds?

From studies in Saanich Inlet, we made a bold move to look at the oceanic environment that came to the coast of British Columbia as part of a trans-Pacific Current. This current originated off the coast of Japan, as a confluence of the Kuroshio and Oyashio Currents. We decided to follow this ocean current backwards, to the coast of Japan, and to see how its biological properties changed in the three years that the waters took to reach our coast. We obtained permission to use the research vessel CNAV *Endeavor* to carry out the first Canadian trans-Pacific expedition. By now, we had a good group of seasoned researchers, including Ed Barraclough, who studied fish, Robin LeBrasseur and Owen Kennedy, who studied zooplankton, Ken Stephens, who was largely responsible for chemical nutrients, and myself, following the phytoplankton. In addition, Dr. Hugh Seki of Japan came with us to study bacteria. Mr. Percy Wicket came as a second-in-charge and helped organize the daily routines, and Dr Kilho Park, a chemist from the United States, was on board to study the partial pressure of carbon dioxide in the different waters of the North Pacific. From Victoria, British Columbia, our destination was Tokyo and Hakodate, Japan. We sailed on March 17, 1969, and Captain MacFarlane managed to tear away a piece of the Esquimalt dock as we steamed out of port. According to Dr. Seki's Japanese folklore, this was a good omen, for it was said that to take a piece of the home port with you meant that you would make it back again!

From the first day out, I wrote a diary of the crossing. When I look back at it, I can relive a voyage that marked an important point in my career. The individual days blend into a total picture of an experience that marked an advance toward studying the whole North Pacific, which would be continued for long after we were gone. My first entry was made as we left home waters, collecting samples of more than sea organisms!

March 18: First samples were collected at 1200 hrs in moderate seas. Most of the scientists were suffering from varying degrees of sea sickness but the situation improved toward noon, and at the evening meal we had a full table. The wind dropped throughout the day, and all twelve stations were taken without major incident. From noon on the 17th March, to 1600 hrs today, we traveled 270 miles. Coming out of Juan de Fuca yesterday evening, the bridge reported two large glass balls floating to starboard—the Captain found it necessary to alter course to retrieve the same (I hope that we are not going to stop for every glass ball from here to Tokyo).

March 19: We have been riding a gentle swell all day, with little indication of the storms which appear to lie all around us. The afternoon was particularly pleasant, with occasional sunny skies and a light fog, against a background of intensely blue water. When the ocean waters are calm, many birds, particularly the albatross, have difficulty flying. They depend so much on the energy of the wind to maintain themselves in the air. Instead, it is a time when they can sit around and preen their feathers.

The midwater trawl was lowered to 300m and brought up a large number of lantern fish, about 30 arrow dragon fish, and a mess of jellyfish, euphausiids, and shrimps—also, one deep sea smelt; anyway, Ed was very pleased with the assortment of material. While this operation was being carried out, the main hydraulic winch started to spout oil; the mechanical defect was not serious but the deck became extremely hazardous to walk on. Apparently, the best treatment for this is powdered brick, which both absorbs the oil and increases the friction under foot. Dr. Seki has rheumatism.

Fire drill was carried out in the afternoon. The life jackets that we put on were the standard issue for the last century—really just a big flotation collar, rock hard, covered in canvas, and stuffed with kapok and cork. I have been told that they are guaranteed to break your neck if you jump into the sea feet first and allow the collar to take the impact of your descent. Fortunately, this was not part of the required drill. A lifeboat was lowered, hoses appeared, bells and whistles were sounded, and then we all went back to work.

March 20: The day started rough, but by noon we had a clear sky and only a moderate swell. During the night, we had to sound our fog horn for about one hour—this is the only disadvantage that I have found to my otherwise luxurious accommodation of the chief scientists' cabin on the main deck (the other scientists, below deck aft, never heard the horn). I talked to the weathership (50°N, 145°W) at about 1100 hrs and arranged for them to take certain samples the following day. Dr. Seki's rheumatism is much better. We have passed through three quite distinct water masses since leaving; for the first day we were in coastal waters, with a shallow mixed layer and high biological production. Tuesday and Wednesday we spent most of the day in transitional water, containing both sub-tropical and sub-arctic species; today we are in clear blue sub-arctic water, with very little biological production (a property which gives the Pacific its blue color), the area of low production started about 100 miles east of the weathership, *Stonetown*, which we passed at approximately 1900 hrs local time. The Captain took us to within about half a mile and it looked as if everybody on the weathership had turned out on deck to see us. The weatherships stay on station for seven weeks at a time, so they were pleased to see us. We exchanged greetings with three blasts on the ship's horn. The captain dropped a mail sack in the sea for them to retrieve, and we quickly disappeared into the night.

March 21: We are quite far in advance of our scheduled position, due to relatively calm weather. Last night, we again experienced a strong swell and the wave height recorder indicated that most of the waves were around 2m, with occasional 4-5m swells. The ship rides very well in this sea and since the swell is head on, the only motion is fore and aft. To date, we have not seen any mammals, except for two porpoises just out of Juan de Fuca. For dinner tonight we had clam chowder, a choice of Alaska black cod, fried sole, or a western sandwich, followed by peaches, and coffee. Everybody's appetite is keen, and some of us require more exercise than food.

Ship's cooks have a very hard time. Either everyone is ravenously hungry, and also willing to complain, if they do not care for the cuisine, or half the personnel are suffering from some degree of seasickness and don't want to even think of food. A good ship's cook has to judge from local conditions whether it is better to prepare a gourmet meal or rehash yesterday's lunch.

By noon today we had traveled a distance of 1040 miles, at an average speed (including three hours of station time per day) of 10.8 knots. Our position is 50°N 150°W, and we have to travel 3333 miles to reach Tokyo Bay.

March 22: Last night, we caught about twenty larval greenling in the midwater trawl—they were alive and in good condition. This is interesting because they must have been spawned at least 300 miles away, which is our closest point to the continental shelf of Alaska. Also, Ed caught a fish which he is not sure about—it is approximately 15cm long, and has scales on it about the size of a fifty-cent piece. There are various ways to exercise oneself on this vessel—the helicopter hangar provides a large covered area in which one can skip (the skippers are Robin and Ed), while the flight deck provides an excellent area for jogging; Kilho Park is a regular jogger, and I enjoy it, if it is not too windy. Dr. Seki practices with his sword in the hangar (preferably when others are absent), and there is also a mat there for those who feel that they must do press-ups or otherwise prostrate themselves. Ken gets his exercise in the chains every day, and Percy sings. This evening, the wind freshened to between 25 and 30 knots, and we had to abandon the evening midwater trawl. Two Laysan albatross followed the vessel in the twilight hours. These birds have white bodies and grey/brown wings and are a much prettier animal than the black-footed albatross we see so often in the Gulf of Alaska.

March 23: We went through our first storm last night, with winds of 30 to 35 knots. The vessel behaved very well and, except for a few belongings scattered around the laboratory, there was no major incident. We appear to be due for a second storm in about 24 hours, with a low of 975mbs. After rounds this morning, the Captain invited the bosun, the first mate, and me in for a drink before lunch. We discussed arrangements for the open house and reception in Tokyo and Hakodate. After lunch, we took the noon station, which lasted for an hour and a half. By about 1700 hrs, it started to blow again, but we decided to go ahead with the midwater trawl that we had abandoned the previous evening. Everything went well, but the wind is coming at us from the starboard, causing us to roll quite heavily. When the bilge keel comes out of the water and catches a wave on the way down, it sounds as if the vessel has hit a log ... I think I'll lie down for a while, it's late and I don't feel like fighting the motion.

March 24: Our fog horn sounded from early morning, until just before noon—every two minutes a ten-second wail. After breakfast, we went over a couple of symposium papers but, unfortunately, the bulb on my projector blew out, which made it very awkward to follow the slides. I never thought of bringing a second bulb, which is surprising, considering how many spares we have for all the scientific equipment. However, I think that we will be fairly well prepared for the meeting with the Oceanographic Society of Japan. Tomorrow, we will hold a meeting in the morning, to discuss the first week's results. Having Percy, Owen, and Doug on watch has been a great help to the rest of us, in relieving us of the more routine aspects of our work. This is especially true in such ordinary things as the correct time, and ship's position—with regard to the latter, the watch have one clock in the dry laboratory that is kept on GMT, and the ship's time, which is always changing, is recorded as plus or minus so many hours GMT—positions are plotted up, so that we don't have to rush up to the bridge every time we make an observation. The fog cleared by afternoon, and we finished the day in bright sunshine.

March 25: Today, we crossed the date line, and it snowed a little. At noon, I received a message from Ted Fleisher (the CNAV representative in Esquimalt), and had an opportunity to tell him a little bit about our trip. There seems to be some bug going around the boat this evening and I have it, so I am going to have a hot shower and turn in early. Vessels are a germ's sweet heaven; there is no escaping a contagious virus, with so many persons confined in such a relatively small space.

March 27: (From Tuesday to Thursday!) We have been riding a severe storm all day, and I believe that the correct term for this sea is "precipitous." The wind has blown 40 to 50 knots, since early a.m., and we have only been making about two or three knots. We were hit by a heavy sea just after lunch, which knocked the door open on the passageway outside my cabin, filling the former with water to the depth of half a meter. It also wrecked all the Van Dorn sampling bottle racks and Ken's continuous temperature and salinity probe, and several pieces of external fire-fighting equipment were sheared off, including a half-inch metal valve. However, the ship appears to ride well, although all is not over, as I write. Certainly, the mechanism by which the power to the propellers is cut off as they come out of the water makes for a much smoother ride than I am used to, in a heavy sea. It is interesting that the crew are quite indignant at my having told Ted Fleisher that we were making fair progress—they believe that the storm is the result. There are, of course, many superstitions held by every generation of sailors ... so never be caught whistling at sea! Right now, I believe that I should not have tempted fate. No more of such reports.

March 28: I forgot to comment on our meeting, at which all the scientists presented their results, to date. This was held in the dry lab at 0930 hrs, on March 25, and lasted until about 1200 hrs. Chemical monitoring has been most successful and with no major breakdowns, we will have a continuous record of nitrate, phosphate, silicate, temperature, salinity, chlorophyll a, and turbidity, from Esquimalt Harbor to Tokyo Bay, and back. We can see from this where the phytoplankton ecology has changed from diatoms to coccolithophorids, where we passed upwelling, and where there has been some production of carbon dioxide (air and sea surface); there are some interesting departures from what one might have expected. To date, the zooplankton sizer has not worked well, but this is only the first time we have tried it, and perhaps we are still too impressed with its ingenuity to delve deeply for the source of the inconsistencies in these data. On the echo sounders, we have passed through some areas with well-developed targets but in general, the 300m scattering layer has been absent on this leg of the trip—young squid and lantern fish continue to be the organisms most often encountered in our hauls, although what one really needs out here is a much larger midwater trawl than we are operating.

March 29: We are running almost two days behind our schedule, due to one bad storm and continuing winds of 25 to 35 knots. This morning, at 0900 hrs, I decided with the Captain not to take the noon station, due to the approach of a low, which was already causing winds of up to 35 knots. However, by 1100 hrs the barometer started to rise, and we had to reconsider our plans. In the meantime, our automatic recorders and the bathythermograph trace clearly showed that for the first time we had moved into waters of the Western Subarctic Gyre. This is the western equivalent of the Alaskan Gyre, but it is a much more biologically productive body of water. It is generally dominated by larger phytoplankton cells, more zooplankton, and a greater abundance of fish.

After discussing the pros and cons of taking our noon day station in moderate seas, we decided that since we should be in the same water mass for the next two or three days, we would sacrifice the three hours of station time to gain some greater distance from the previous noon station, only 160 miles astern (compared with an average interval of 265 miles). Instead, we spent most of the day tidying up our data and repairing equipment damaged in the storm two days ago. By evening, we had crossed two weather fronts, and instead of encountering the storm, we came out into a high pressure area with the calmest water we have seen for some days. The Captain has put our second engine to work for the first time on this voyage, and we are buffeting along at 16 knots, with lots of white water coming over the bow. There was a movie on this evening called *The Way It Is*.

March 30: We have been at sea for nearly two weeks, and we are about 1,600 miles from Tokyo. Today, the sea was very calm and we had two engines running all day—we should cover our greatest distance in any one day. Our daily schedule consists of breakfast at 0730; start of daylight observation program at 0800 (there are actually twelve stations per day, staring at 0200 with the night watch); lunch at 1130; the ship stops at apparent noon and we collect data from depth profiles for about an hour and a half, coffee break at 1430; dinner at 1630; and one hour evening station at approximately 1930, which is primarily for collecting animals with the midwater trawl. In the evening, there is usually a movie, or perhaps a card game, and then most of us turn in by 2300—except the watch, who carry on in shifts sampling and checking recorders through the night. It has been a pleasant day.

March 31: Last night, Ed caught two adult fish in his trawl that were transparent, like herring larvae. Transparent fish are known to occur in the Antarctic, where it is reported that the lack of a need for hemoglobin in very cold waters has allowed for the evolution of a few such species—however, as can be seen, we are not very well up on this subject, since these animals do not appear in any of our traveling library books. Perhaps it is a first, but if not, at least it will be the first time that this fish has been reported to occur off the coast of British Columbia (at approximately 45°N and 158°E!). We had a successful morning, but by the afternoon the wind had started to blow, and by evening we had to heave to in heavy seas, with winds of 40 plus knots. We are about 1,000 miles out of Tokyo, with good hopes that we will make it at least by Easter Day.

April 1: Still a strong wind of around 30 knots and we have been hove to all night. This morning, the skipper decided to run with the wind for a while—this will take us due south and, hopefully, into a high pressure area. Either way, we will be making some distance to Tokyo, although it will not be on the great circle route, which we have been following to date. One interesting aspect of our work has been the use of the 200kh Furuno sounder, in combination with the Longhurst recorder. Without going too far into details, the former constantly shows a shallow scattering layer at about 40m. This, we can describe in detail, from collections made with the Longhurst recorder, which takes discrete depth samples; the result of this is that we find the zooplankton concentrated in very narrow zones—a result that we

The Sea's Enthrall: Memoirs of an Oceanographer

cannot obtain with the conventional zooplankton net. The zooplankton layer is well above the deep scattering layer, which is usually composed of fish and squid. The concentrations of zooplankton are comparable with those found in the Strait of Georgia, which conventional wisdom would classify as being much more productive than the open oceans. It has been said that plankton is so thinly dispersed that it would always be uneconomical to collect it—plankton *dispersion* in the oceans has never been studied on this scale, and with the kind of apparatus which we are using today. I hope that Ed, Robin, and Owen get a note off about this aspect of their data (they did later, and it was published in *Science).* The sea was much calmer by noon, and we are now back to around 14 knots in a small high pressure area. The snowfalls, which we have experienced several times in the last few days, have changed to rain squalls, and the sea temperature has gone from 2.8°C to nearly 10°C in the last forty-eight hours.

We have passed through two major current systems in the last three days. After crossing the Western Gyre, we entered the Oyashio Current, which is a cold current coming from the north. Then, with the change in water temperature, we started to enter the Kuroshio Current. This current comes up from the South China Sea and is generally less productive than the Oyashio. It is the fastest of the world's major currents, with speeds of up to 7 km/hr, which is faster than most people can run. The two currents meet at approximately the latitude of Tokyo, and then head east across the Pacific, arriving at the coast of British Columbia about three years later. Occasionally, on the coast of British Columbia, we find giant leatherback turtles that have drifted all the way across the Pacific from their home in Malaysia, carried by the northward-flowing Kuroshio Current. Another animal known to be caught up in this system is the great white shark; one of these was washed ashore on the Queen Charlotte Islands several years ago. The junction of the cold arctic waters and the warm tropical waters off the coast of Tokyo results in highly productive water, and fisheries for sardine, mackerel, and anchovy. A similar area is the junction of the cold Labrador Current and the Gulf Stream, off Cape Hatteras, at the same latitude in the Atlantic Ocean. Both regions are renowned for their storm activity.

April 2: We are running with the wind in a moderate sea. I must say that I have great respect for Captain MacFarlane's ability at sea—with years of experience derived from cruises covering an area between Aden and

Shanghai, he has a great sense for when to heave to and when to make time. I don't think that I would ever care to have an experienced captain replaced by a computer. The wind freshened in the evening, and we were making a little progress by nightfall.

April 3: Our most frustrating day to date. We are about 450 miles from Tokyo and the sea is very rough. We are unable to take our regular station and unable to make any headway toward Tokyo; we had hoped to dock by Friday, but as long as this storm lasts, we may even miss our scheduled meetings with the Oceanographic Society of Japan. Later—the Captain started to take the boat across the wind at about 1500 hrs this afternoon, and although we have been rolling heavily, we have been making some progress and the sea is becoming calmer. We decided to go ahead with the evening station and so made our first collection in the warm waters of the Kuroshio Current—the thermocline is isothermal at 18.1°C to 90m.

April 4: Our ETA Tokyo is now 1100 hrs, Saturday 5 April—after three weeks rolling around the Pacific, this is most welcome news. I must start addressing envelopes.

April 5: Shortly after cabling our ETA, the wind came up and we have been sitting through the worst storm that most of us have experienced. The wind speed at 0300 hrs was 60 knots, gusting to 70, and the wave height recorder showed 10-15m waves, with one wave off the scale at about 20m. It is a very frustrating because we are so close to cover but unable to make any progress in this sea. At about 1100 hrs this morning we received a weather forecast which warned of a northerly monsoon coming up from Taiwan—I had hoped that we were too early for the monsoon season. The Japanese have a special name for this wind—it is called "Taiwan Bouzu" (Bouzu, in Japanese, means "gangster"). According to Dr. Seki, this is *very terrible wind,* and we must seek shelter—but where to go in a typhoon? At about the same time as we received this forecast, the wind went about nearly forty-five degrees—I wish you could have seen the 15m swells when the 60-knot wind suddenly started to blow from another direction—the top of every wave disintegrated into a cloud, then a little sunlight broke through, and the sight of blue Pacific water, laced with white foam and threaded with streaks of spray, was ecstatically beautiful. This situation lasted for about one hour and then, to our great relief, we came into a calm sea. This was the eye of the typhoon, and we quickly took advantage of the relative calm to head into port.

April 6: This is Easter Day, the time is 1200 noon, and we are just coming up to the Tokyo pilot station. We will tie up at 1500 hrs, and so start our visit to this sophisticated maritime nation. Everyone deserves a few days ashore, and I'm very glad that we escaped from the "Taiwan Gangster."

One slight glitch in our arrival in Tokyo was that the captain had radioed for an escort tug to meet us at the pilot station and guide us to our berth in that busy harbor. The message had been sent before the last storm, when our ETA had been about 48 hours earlier. The tug had been standing-by ever since, and the docking bill came to $70,000 for this service. I am not sure how this sum was settled, but I remember the captain being very alarmed at this potential expense.

In the next few days, we gave lectures at the meeting of the Oceanographic Society of Japan, and we held an Open House for the public on another day. We were magnificently entertained by the Japanese in Tokyo, especially by Professor Uda, who arranged for us to be honored guests at the annual meeting of the Oceanographic Society of Japan. He also took us on a tour of the Tokyo fish market at 5 a.m., when the selling activity of fish from all over the world is at its height. This is probably the biggest fish market in the world. The dominant species seems to be tuna, and a large blue fin (possibly caught off Nova Scotia) can sell for $100,000—that's a lot of sushi. Every fish imaginable, from the poisonous fugu (which must be cleaned and handled by specially licensed fishmongers), to boxes of shellfish, and sleek yellowtail raised in aquaculture, come under the auctioneer's hammer. Bidding is brisk and furious, with no bystander having a clue as to who is the winner on any particular bid. Unlike many fish markets that I have visited, the Tokyo market is remarkably free of the smell of fish, which indicates that nothing has been allowed to spoil on its way to market. The market employed a Japanese expert, the late Professor Abe, to identify, as to species, any catch of fish that had not been previously reported. This was necessary, in order to eliminate possibly toxic species.

We traveled north to Hakodate after a few days in Tokyo, and were again warmly entertained by Professor Motoda, the senior planktologist at the University of Hokkaido. The people of Hokkaido are less formal than the inhabitants of Tokyo and the party that we were given was full of spontaneous events, such as everyone having to stand up and sing at least one song. Dr. Seki's mother booked a one-night stay for three of us at a *ryokan*, a traditional Japanese inn. We were met at the door by the owner and three charming hostesses, who instructed us with sign language where to bathe, sleep, and eat. The bathing was in a hot spring. We were seated on little wooden stools and had to scrub ourselves before entering the hot geothermal pool. The temperature of the pool was enough to take your breath away, and we all emerged looking like boiled lobsters.

We left Hakodate on April 14, with a large crowd of Japanese wishing us goodbye, and followed the same routine of sampling on our way back to Canada. When we arrived back in Esquimalt, Canada Customs had set up a temporary shed on the dock where we tied up. We all had to appear before them and show them what we had brought back. As the senior scientist on the vessel, I had been given a large number of gifts by the Japanese, and so my 'purchases' seemed excessive. I gave some away to other scientists and presented the ship with one particular souvenir. Nevertheless, I was still left with a bill to pay because of the rather strict customs duties at that time. A search of the boat was also conducted, presumably for any contraband. Oceanographic vessels had been known to smuggle goods, and one on the east coast had been caught with a large amount of hashish in jars neatly labeled as plankton samples. However, we came away clean and were allowed to leave the dock.

Our trans-Pacific cruise gave rise to a new program in cooperation with Dr. Andersen at the University of Washington. This program entailed using commercial ships to collect data throughout the year, over approximately the same transect that we had followed. The data collected from such cruises eventually supplied some of the knowledge about how the whole North Pacific could change its ecology—a process that became known as a "regime shift," and which was shown later, by other scientists, to greatly affect the abundance of salmon in the area.

In the latter part of 1969, I received a phone call from Dr. Charlie Goldman asking me if I would stand for election as President of the American Society of Limnology and Oceanography (ASLO). This was, and probably still is, the leading society for aquatic scientists in the world, except that it does not contain many physicists.

"I don't think I'm ready for that just yet, Charlie," I replied.

The Sea's Enthrall: Memoirs of an Oceanographer

"That doesn't matter," he said. "The executive has selected Liugi Provasoli. He's sure to get the job, but we have to put someone else on the ballot."

Apparently, I was seen as a reasonable, although unlikely to win, contender. Believing I had no chance of being elected, I said that he could put my name down as one of two candidates. A few months later Charlie phoned again with the surprising news.

"Congratulations," he said. "You have been elected President of ASLO."

For some reason, the membership had not voted overwhelmingly for the older and much more established candidate. The Presidency was a prestigious position, and my election to it resulted in my being invited to give a number of lectures at institutions, where I was surreptitiously viewed as potential material for their staff. When I went to Columbia University in New York, for example, the Dean, who was over-reaching himself, asked me, "So why do you want the job?"

I said, "I didn't know there was any job and if there is, I don't want it. I have only come to give a couple of lectures and talk to some students at the university's invitation."

I am afraid that my term as president of ASLO was not spectacular, although membership did increase substantially that year. I was ill-prepared for the job but I made up for this later, when I was chairman of the organizing committee for the annual ASLO conference, held at the University of British Columbia (UBC). This was a very successful meeting and ended up by netting the society about $10,000, instead of costing them anything.

At the end of 1969, Dr. Seki, who had been visiting our group as a post-doctoral fellow, went back to Japan. We were all very sorry to see him go. I had first invited him to study with us when I went to the Pacific Science Congress in Tokyo, two years earlier. He had published some interesting work on marine bacteriology, and we needed his talents in our group. When he arrived in Canada, he brought his traditional Japanese sword with him on the plane. Unfortunately, he left it on the plane after I met him and we had to go back to the airport to retrieve it.

"No problem," said the customs inspector, "it was turned in about an hour ago and I am glad we can return it to the owner."

This was not the kind of reaction one would get on this subject from airport officials today! After his return to Japan, Hugh Seki became a lifelong friend and we have visited with each other over the past 30 years. His family came from the Samurai class of Japanese warrior, and he was an expert in

Kendo and Aikido. He represented the advanced state of Japanese oceanographic research and was the first of a large number of friends I made in Japan over the years. Getting to know people from a very different culture allowed me to recognize how rich life could be, however ordinary one's beginnings.

While the Japanese people have exported their industrial goods, the unique traditions of Japan, including Shinto shrines, the Kabuki Theatre, Japanese gardens, sumo wrestling, their royal family, and a host of other traditions, have remained strictly Japanese and largely not for export. While the people have embraced western mannerisms for business and science, they have managed to maintain their culture, despite the growing uniformity of a world dominated by Hilton Hotels, international sport, stock markets, airports, and thousands of miles of paved highways. Dr. Seki's visit with us had taught us much about Japan, and he was soon followed by another Japanese scientist, Dr. Takahashi, who started to work with our group on a new research project.

It was 1969, and this year marked another turning point in our research as a team of biological oceanographers. We decided to apply our work to a long standing problem in aquatic science—how to enrich salmon-producing lakes, in order to enhance salmon production. Although our experience to date had been mostly marine, I had been looking for some project where we could demonstrate our knowledge in a practical way. I approached the task with great enthusiasm. Our team was supported by some outside scientists, while others were openly hostile to our plans. The reason for this hostility varied, but I think we were seen as infringing on the activities of scientists who had been hired as limnologists, while we were oceanographers. This attitude was like a mini turf war, but our claim to undertake this project was based both on our team approach and on our scientific experience, and not on wanting to take away another scientist's budget.

In general, we planned to try our scientific skills on Great Central Lake, a large lake in the middle of Vancouver Island. It is about 35km long, but less than 2km wide in most places, and about 400m deep. After working out some scientific details, our plan was to add 100 tons of nitrates and phosphates to this lake, which was considered to be ultra-oligotrophic, meaning that, in fact, it was rather like distilled water, as far as being able to grow any plankton was concerned. Nevertheless, sockeye salmon used this lake as a nursery area for their young, although the young fish that emerged from the lake into the sea were the smallest and scrawniest fish on the whole coast. If we had taken the traditional population dynamic approach to this lake, we could argue that if we could kill off some of the very numerous small fish in the lake, then the ones that survived would grow bigger because there would be more food per fish. However, we took what we had learned about the productivity of aquatic systems, and instead, decided to add 100 tons of inorganic fertilizer in 10-ton lots, spread over ten weeks. By doing this, we had reason to believe that we could raise the total productivity of the lake, so that all the young salmon would grow larger and survive better when they reached the sea, to give an increased return of adults.

There was much logistical work to be done, but at first, our problem was more political than either logistical or scientific. Jack Davis, the then Minister of the Environment, had started a program back east on the Great Lakes that was designed to keep nutrients out of lake waters, and here were we advocating putting nutrients into a lake. He was not the most pleasant man to deal with, but we managed to get the backing of a very distinguished Fisheries Research Board scientist, Dr. W. E. Ricker, and with him on our side, the problem of Jack Davis's lack of understanding between eutrophic (too much nutrient) and oligotrophic (too little nutrient) disappeared.

There was one more hurdle, however, and this was the provincial Pollution Control Board, whose members pointed out that the lake in question was also the water supply for about 5,000 people in the city of Port Alberni. We tackled this problem by advertising our intent in the newspaper, and followed up with a press conference for the people of Port Alberni. At this presentation, I drank some of the water which I had laced with nitrates and phosphates in the same concentration as we were to use in the experiment. Actually, this was unnecessarily dramatic because these nutrients would be taken out by the algae long before the water was used as a domestic water supply. However, we won the day, and fertilization of the lake started in 1969.

I have retained a complete file on all the exchanges we had with the public and with other scientists on our planned experiment; in addition, the results of the whole experiment have been published. To the bitter end and beyond, some scientists at the Biological Station refused to believe in our work and in fact, after I left the station for the university, a fisheries scientist was hired in Nanaimo specifically to challenge our claims. It was surprising because the claims were self-evident.

First of all, we could detect increased production in the lake. Secondly, the young fish grew about 35 percent bigger, and finally, after three years, when they returned to the lake as adults, the numbers of fish returning was increased on the average from 52,000 to 373,000—a sevenfold increase in salmon production! At the same time, the benefit to cost ratio of the treatment was a healthy 3:1. It is difficult to imagine what other single event could have produced such a magnificent result and indeed, in spite of the skeptics, the fertilization of all salmon-producing oligotrophic lakes was adopted as part of the salmon enhancement plan from that time onwards.

We had demonstrated that the environment was important in the rearing of fish. The real opposition to our experiment lay in the deeply entrenched, but largely ineffective, ideas that had pervaded fisheries management for decades. When there is a whole school or generation of scientists believing one thing that they were taught in college, it is surprising how difficult it is to change their thinking. The fisheries of the world have suffered greatly from the early beliefs of many fisheries scientists. Change is an integral part of life, but as people age, they sometimes feel threatened by new ideas that challenge the *status quo*.

In 1972, an opportunity arose for me to apply to the University of British Columbia for a tenure-track position in the Department of Oceanography. My application was successful and I was appointed as a professor jointly in the Departments of Oceanography and Zoology. My desire to change from government to university employment was motivated by my need to write a textbook that would deal with biological oceanographic processes in the sea, and how they could be used for fisheries management, and aid in understanding the problems of pollution, which had recently become of interest to governments. Writing textbooks was not really a function of a government scientist but of someone who was teaching students. This was particularly true for the teaching of graduate students, because they had already decided to have a specific interest in the field of oceanography, but there were surprisingly few books available that were committed to studies in this subject.

My thinking at this time was also influenced by an American scientist, Dr. Carlton Ray. He had written a paper in 1970, in which he described what he called the "Marine Revolution." Essentially, he defined the need for biological oceanographers to do a lot more than peer down microscopes and describe different species of bugs in the ocean. The national economy and legal complications were starting to require answers to some important questions on marine pollution, world fisheries, and the establishment of 200-mile economic zones. As a result of his opinions, I saw a need to include practical aspects of ocean science in any textbook on biological oceanography.

Captain of My Own Boat

VIII
Free to Chart an Independent Course

1. Getting things shipshape

Compared with my employment by the federal government or UNESCO, I was now appointed to a position with complete freedom. My choice of research project was contingent only on my ability to obtain funding. Of course, there was a university commitment to teaching, and a rather looser responsibility to engage in administrative activities. I was just as keen to start teaching as I was to write a textbook. I started four new courses; one general introduction to biological oceanography, one advanced course on ecosystems for graduate students, one on marine pollution, and a methods course on measuring biological and chemical parameters. Strangely, none of these courses existed in the department, which did not teach any undergraduate courses, and only had a rather irregular graduate teaching program in biology before my arrival. The department was strong on physical oceanography and had some chemistry courses. I felt that the introduction of my four courses was a healthy start in academia.

The most difficult of these courses to teach was the one on marine pollution. A group of undergraduate engineers decided to take it. I was not sure why, because their questions and attitudes were often openly hostile to the content. For one thing, at that time, engineers favored long pipes, whether up in the air (chimneys) or along the ground (sewage disposal), as the answer to sundry pollution problems; the longer the better. Ideas about trace amounts of pollutants being a safety risk did not appeal to them. Some of the students would show their disdain by reading newspapers while I was lecturing. Since the exam material was more a gauge of opinions than a regular learning exercise, perhaps they thought to gain more from the press than from my class. The course was, nevertheless, a success for others, and was continued after I left.

After organizing my teaching, my next task was to find funding for my own research efforts. Administration did not interest me, although it certainly interested some academics who wanted to climb the ladder toward department chairs, deans, and beyond. I applied to the National Science and Engineering Research Council (NSERC) for money to study marine ecosystems.

My application was rather vague because I did not couch it in terms of the favored theory of scientific research, which began with the need for a hypothesis. In 1934, Sir Karl Popper, then Dr. Popper, who was a key figure in the field of research theories, explored this approach as it applied to science, in his definitive book *Logik der Forschung* (*The Logic of Scientific Discovery*). He maintained that science progresses through a testable hypothesis, beginning with an idea about how something works, based on available data, then proceeding to demonstrate the degree to which that hypothesis is correct. Many challenged this approach, one of the most forceful being Thomas Kuhn, proponent of the "paradigm shift" concept. In 1970, Kuhn rebutted Popper, claiming that much of scientific knowledge actually comes about as the result of astute observation, not from hypothesis testing. His statement on this is: "Unanticipated novelty, the new discovery, can emerge only to the extent that [the scientist's] views about nature and his instruments prove wrong" (from *The Structure of Scientific Revolutions*, U of Chicago P, 1970).

In other words, as Isaac Asimov has said, the exclamation often voiced by a scientist when looking over his data is not "Eureka!" (I have found it), but, "That's funny ... ." Thus it is the anomaly in the data that emerges as the point of original discovery because it differs from what the scientist thought would happen, based on accepted scientific theory. Unfortunately, it was Popper's book that was widely read by funding agencies and not so much the works of Kuhn.

This "hypothesis testing requirement" of funding agencies was detrimental to those who, like myself, often inclined towards exploratory research. I, and others, thought that Nature could lead a researcher into extraordinary problems, many of which might not be included in a previous hypothesis, because they were not known about when the grant applicant filed his request. It is a sort of "catch 22" situation. A simple example of an unplanned discovery was Fleming's discovery of penicillin, which opened up the whole science of antibiotics. Another, related to marine science, was the discovery of biological communities in volcanic vents under the ocean. No hypothesis proposed their presence. They were discovered by accident, when a submersible vessel stumbled across some temperature anomalies in the deep waters of the South Pacific (cf. Edmond, J. M. in *Oceanus*, Vol. 25, 1982). Nature is full of such surprises. I like to go into a general area (in my case, marine ecosystems), see what I can find out, and be ready to change direction "with the wind." Not all rescarch can be conducted in this fashion, and certainly our work on salmon in Great Central Lake was very directed. However, now that I was at the university, I felt that this was an opportunity for exploratory research.

The second driving force for my research was the need for some kind of methodology (technique or instrument) that would give me unique results. This, again, was not conventional science, which tends to frown on instrument research and sees it more as being engineering than science. Ironically, at least half of the Nobel Prize winners actually achieved their breakthroughs by using a unique method they developed to measure something completely new. Unfortunately, the NSERC committee awarding funds became increasingly hypothesis minded, while my research interests were going in a different direction. Thus, it could be difficult for me to be funded. To be original, however, a scientist has to expect to be in the minority. Perhaps, by having a minority approach to the philosophy of research, I could also assure the originality of my findings. This funding problem was not so much an impediment at the beginning of my research at the university, but it became a major issue later on, when I found that my research grants were eventually decreased.

Mother (1948)

Christopher (left), me, and Mother
(Port Said, Egypt, 1936)

Grandfather, at 85 years of age (1945)

My brother, Chris (right), and me (Bude, 1943)

"Bobsy" (left) and Chris (Bude, 1944)

Christ's Hospital dining hall, and the water tower I climbed with my box camera (1948). *Left*: The picture I took!

Playing the bass drum in the school band, St. Matthew's Day Parade (London, 1948)

Wearing the Christ's Hospital uniform (1949)

Swimming champion, Macdonald College (1953)

My "second love" – sailing on the Lake of Two Mountains (Quebec, 1955)

Graduate studies at McGill University (1956)

"I want to get down!" (Cairo, 1963)

Scientists on board the CNAV *Endeavour*, en route to Japan (1969)
(*Back row*: Wicket, Kennedy, me, Stephens, Seki. *Front row*: Hager, Robinson)

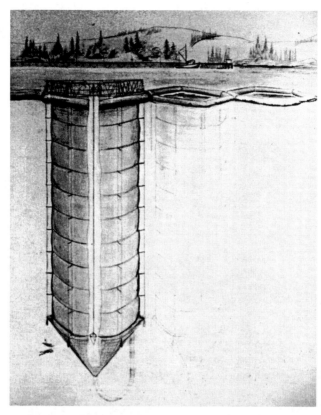

Artist's rendition of a mesocosm
(SCUBA divers drawn to scale at lower left of bag)

Mesocosms in Saanich Inlet, BC (1974)

Preparing for a hike with students (Black Tusk, BC, 1979)
(*Standing*: Fong, Perry, Lalli, Sharp, Parslow, me. *Kneeling*: Carruthers)

Mesocosm research scientists in Xiamen (1985)
(*Front Row*: Profs. Li Guanguo, Harrison, C.S. Wong, me, unidentified, Prof. Wu Baoling)

Preparing to visit a drilling platform in the Arctic Ocean,
Tuktoyuktuk air base (Winter, 1988)

Department of Fisheries and Oceans research vessel, the *John P. Tully*,
in the Arctic Ocean (1990)

Playing tennis in Kaohsiung (Taiwan, 1993)

Authors of *A Manual of Seawater Analysis* (1985)
(Me, with Drs. Maita and Lalli, 1994)

Receiving The Japan Prize (Tokyo, 2001)

Entering the banquet room, following The Japan Prize ceremony (Their Majesties, the Emperor and Empress, followed by the two laureates and their wives)

Sightseeing with my wife, Carol, after The Japan Prize ceremonies (Tokyo, 2001)

Honorary DSc ceremony, with my daughter, Dr. Allison Krause
(University of British Columbia, 2002)

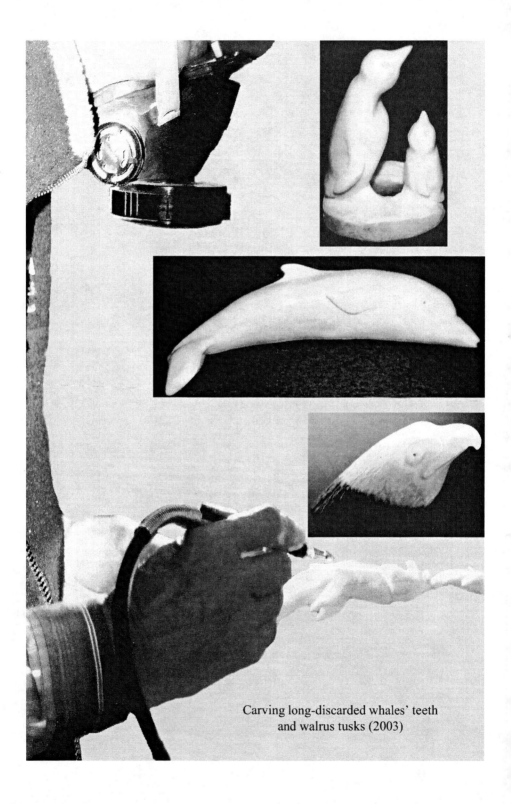

Carving long-discarded whales' teeth and walrus tusks (2003)

During this change over from government to university, I was elected to be president of the International Association of Biological Oceanographers (IABO). This association is a member of the International Council of Scientific Unions (ICSU), which represents science outside of all government control. The ICSU (and consequently, the IABO) express opinions on science that are free from political or economic interference. In general, the two organizations are highly respected for their independent opinions. As the president of IABO, I was invited to annual meetings of the Scientific Committee on Oceanic Research (SCOR), which is another body belonging to ICSU, and which organizes international oceanic research. Despite their dull sounding names, these interagency connections were very important to me in making contacts with other oceanographers in the world. In some cases, they provided me with opportunities to take part in international exercises.

One such exercise was to standardize the method for measuring chlorophyll in the sea. The measurement of chlorophyll in seawater is an important parameter for telling how much planktonic plant life there is in any particular area. If we did not have a standard method for this compound, the results obtained by oceanographers in different countries would be meaningless, in a global sense. I was asked to chair such a meeting of scientists, and we published our recommendations as the first volume in a new UNESCO series on oceanographic methodology.

SCOR held annual meetings of the executive in various parts of the world. I attended these as an ex-officio member. We always tried to involve the national members of SCOR in the country in which we had our meetings. This helped to affirm the international nature of ocean science. The meetings were heavily dominated by physical oceanographic science, but this was fine with me because we needed a well-developed understanding of the physics of water movement in the ocean, before we could attempt to understand much of the biology. My role as the only biologist member of this committee might have seemed to some to have been entirely passive—like that of a 'sleeping partner' in some big business. However, I felt that we were not ready for world-scale ocean biology, so long as we lacked an understanding of such physically gripping events as El Niño—the sudden flow of massive amounts of warm water from west to east in the tropical Pacific. At one of the meetings, someone referred to a curiosity of Nature in which a certain species of red crab on the coast of Peru always disappeared long before the arrival of an El Niño. Perhaps the red crab should have been invited to some of our SCOR meetings, instead of me.

One of my more constructive activities with SCOR was to persuade them that we should have a working group on the effects of fishing on the ecology of the ocean. I had written some articles on this subject because it seemed to have been largely ignored in the rush to manage fisheries, rather than to understand ocean ecology. This effort was successful, and eventually led to more than one symposium on the subject.

At our meeting in Sao Paulo, Dr. Federov, who had been my boss at UNESCO, had become president of SCOR. The conventional dress for such professional meetings had recently changed in the west, from suits and ties to a more informal dress code. Konstantine Federov arrived in a brand new Italian silk suit and tie.

"Just when I have been able to afford a decent suit for these meetings, everyone turns up in jeans and sweaters!" he said, wryly grinning. At home, in the economic climate of the Soviet Union, his suit had given him status, but his appearance was an oddity on the international oceanographer circuit.

These meetings of SCOR coincided with my arrival on the UBC campus, where I had formulated some definite ideas. For some time I had thought that oceanography, as a science, should be located at the University of Victoria and not in Vancouver, at UBC. There were some good reasons for this. One, which I advanced most strongly, was that the new government Institute of Ocean Sciences was being built near Victoria. Since we needed to cooperate with this organization for the use of ships and some other expensive equipment, we would be better off to be near them. A second reason was that UBC was a very large campus, with enormous departments of Chemistry, Zoology, and Geology, as well as large faculties of Law and Medicine. With only about ten faculty positions, oceanography was a very small frog in a very large pond. Neither of these problems were unique to our university, and the same debate regarding location has occurred in other countries.

With our director's permission, I went with two other faculty members to discuss this idea with the Dean of Graduate Studies at the University of Victoria. He was very sympathetic to the idea and assured us that oceanography would be a star attraction for their campus, if we could all transfer as a unit. Our director of Oceanography was not inspired by the idea. Having spent almost twenty years as Head, he was not about to move. The democratic process of voting as a group on the idea did not give us any clear majority, many people being more worried about changing their domestic security for something new on Vancouver Island. It was all a wasted effort on my part, but, ironically, within about twenty years, Oceanography at UBC ceased to exist as a department and was amalgamated with Geophysics and Geology. At the same time, on the University of Victoria campus, a strong institute of ocean studies has been established. As in many universities in North America, it has become just too expensive for young professors to find accommodation in big cities, like Vancouver, and our move to Victoria would have solved this problem also, because small campuses are usually surrounded by cheaper housing.

2. Saving the oceans

In the early 1970s, the National Science Foundation in the United States undertook to investigate marine pollution. This was part of the activities contributing to the UN's International Decade of Ocean Exploration. At a conference at FAO in Rome, in 1970, a large number of marine scientists presented papers on marine pollution. It seemed that pollutants were fairly widespread in traceable amounts, but no one really knew what effect they might be having on marine life. There were still many unknowns. It was a great achievement of the chemists to be able to measure a compound in parts per trillion, but what did it mean, ecologically? DDT, for example, which was widely used in tropical countries against the malaria-carrying mosquitoes, occurred in ocean seawater at concentrations ranging from about 1-10 parts per trillion, and at a thousand times this level in some coastal waters. Was this level of DDT harmful to marine organisms? It was known that DDT could be concentrated by some marine organisms to levels that were thousands of times more concentrated than in seawater. How dangerous was it to have 1 or 2 parts per million in fish?

We could approach this problem two ways—either individually test thousands of organisms from plankton to fish, as to whether their survival was changed in any way by these concentrations of DDT, or, test a whole ecosystem and see how it reacted to the same levels of DDT. If we took the first approach, it would be very expensive, and in addition, we would never know if all the important organisms had been tested. By taking the ecosystem approach, however, we had a very large number of organisms present and the question was whether the whole ecosystem changed its function under the stress of very low levels of pesticide, or of any other pollutant. The National Science Foundation decided to fund the second approach, and in order to perform our task, we recreated the concept of the large plastic bag experiment, with which I had been associated much earlier, with Dr. Strickland.

The program was called CEPEX, which stood for Controlled Ecosystem Pollution Experiments. It was funded by the National Science Foundation to the extent of about a million dollars per year, but since it was declared to be an international program, additional funding came from the participation of German, British, Japanese, and Canadian scientists. The site chosen for the program was Saanich Inlet, which was the same area my group had studied a decade earlier. Our knowledge of the area was part of the reason for choosing this site. A committee of four persons, two Americans, one Englishman and myself, organized this program.

Once again, before we could start, the overly zealous Minister of the Environment Jack Davis intervened to tell us there was no way we would be allowed to pollute Saanich Inlet. While he did not understand our experiments, he forbade us to carry them out. Of course, we had no intention of polluting Saanich Inlet, but if publicity-minded Jack could score a point for watchfulness, here was a repeat of the opportunity in Central Lake, a few years earlier. This was a problem, but because Davis was always impressed with the opinions of more senior persons, this time we obtained the support of Dr. Bob Stewart, who was Director of the new Institute of Ocean Sciences in Patricia Bay, on the shores of Saanich Inlet. He explained the logic of our experiments to the skeptical Minister of the Environment, who then let us carry on.

The large plastic bags were now called "mesocosms," and they were made (by Calgary Tent and Awning, Alberta, Canada) to contain up to 2,000 tons of seawater. They were open at the top and descended to a depth of 10-20m. Water was trapped in these containers when the edges were raised by divers, who then attached the whole mesocosm to a large doughnut-shaped float of 5m or 10m in diameter. We carried out the experiments by having one control mesocosm and then adding, for example, 1 and 10 parts per billion of a pollutant (e.g. copper) to two other mesocosms. We then followed the whole pelagic environment in each container for about a month, to study changes in the levels of phytoplankton, zooplankton, and fish. Up to forty scientists and technicians of different nationalities might be involved in such an experiment.

Meanwhile, the press were having a field day with wild stories about dying oceans and marine catastrophes. Such scare tactics sold papers and kept publishers happy. Even eminent explorers, such as Thor Heyerdahl, were caught up in misplaced fervor to save the water of the planet. His 1975 *Saturday Review* article, "How to Kill an Ocean," contained inaccuracies from the simple reporting of 1,700m as the average depth of the oceans, when it is closer to 4,000m, to a quotation that we depend on plankton for "the very air we breathe." The latter had been picked up from another publication (*Bioscience*, Vol. 18, 1968), which said that the oceans were the main supply of oxygen to the atmosphere. This mistaken idea had been corrected twice, in volumes of *Science* (Vol.168, 1970) and *Nature* (Vol. 227, 1970), but the error was then repeated in the *Saturday Review* article five years later! The reporting on what was happening to the oceans was out of touch with reality.

It was in this context of unrealistic and irresponsible reporting that our findings were all the more significant. We began to publish our results from CEPEX. I suppose that the biggest overall finding of our experiments was that the ocean ecosystems we studied did not "die"; there was no "catastrophe"—words frequently used by the press—but there were often some subtle changes in the ecosystem which sent it in a new direction. For example, in the presence of 10 parts per billion of copper, one kind of phytoplankton died off, to be replaced by another, quite different type of phytoplankton, essentially with no change in the amount of photosynthesis. The message from this kind of experiment was that the oceans were very resilient and had evolved to compensate for such changes. After all, ocean ecology in the Post-Cambrian, about 500 million years ago, had been subject to many pollutants from volcanic emissions, and if ocean life was so very sensitive to such pollutants, it would never have evolved at all.

Perhaps Shakespeare had it right in the lines: "Nature with a beauteous wall doth oft close in pollution" (*Twelfth Night*, Act I, Scene 1). In the days of the Bard, sewage was allowed to drain from houses into ditches. The magnificent growth of plant life around these ditches ('a beauteous wall') was what, I imagine, attracted his attention.

The consequences of environmental impacts on the oceans became the subject of a large number of quasi-scientific organizations. These organizations can be listed in terms of their militancy towards the establishment. Some were more militant than others, such as Greenpeace, closely followed by Friends of the Earth. Then there were such organizations as the Sierra Club, which depended in part on subscriptions, and World Wildlife, followed by the Nature Conservancy. Finally, there were organizations created as environmental companies (e.g. Rescan), and the government agencies, such as the Environmental Protection Agency (EPA). Sometimes I strongly opposed the views of the more militant of these organizations, while at other times, I supported them.

In general, the need for funding often decided how extreme would be the attack on a particular issue in the marine habitat. Militancy was in direct proportion to the need for funds. Since, at the bottom of the list, government and company environmentalists were supported directly or indirectly by the tax payer, they tended to be rather quiet about their activities. These were often the most reliable protagonists because they had the best data. Greenpeace needed public donations and therefore often pulled off spectacular protests. The sailing of the Greenpeace boat into the zone of French nuclear tests was a newsworthy event with which I sympathized.

On the other hand, another organization, The International Fund for Animal Welfare, an agency funded by appeals, used a picture of a baby harp seal as a symbol of what we must save in our environment. The harp seal is a superabundant species, second only to the crabeater seal in the Antarctic, in terms of numbers. It was the basis for a local Newfoundland fishery, and not an endangered species, but it does make a good poster for public support. The Nile crocodile is a threatened species. However, no one who was fund raising would use this animal as a symbol.

Oil spills attract immediate press attention and all sorts of environmental groups jump on this bandwagon because even a single oiled bird is a great photo opportunity. In fact, oil spills are deadly over a limited area but are not catastrophic to ocean ecology. Oil has been seeping into the oceans for millions of years and this is a natural process. In the case of the Valdez spill in Alaska, the compensation given to some fishermen was greatly in excess of the damage done. In contrast, there has been practically no challenge by the environmentalists to the massive slaughter of fish in the oceans by the fishing industry. This is by far the greatest anthropogenic impact on the oceans and, except for some targeted by-catch species, such as porpoises, turtles and birds, there has been no outcry, until relatively recently, about the massive overfishing that has occurred throughout the world's oceans.

The CEPEX program lasted for about seven years, from 1972 to 1979, although by the end we had established satellite facilities in other countries, and the whole idea of mesocosm research had been fairly well accepted as ongoing science. Countries that were now carrying out mesocosm experiments included Germany, France, Belgium, the United Kingdom, and the United States. The UK facility was located in Loch Ewe, Scotland, and the US facility at the University of Rhode Island. The British version was very similar in design to the Saanich Inlet mesocosms, but the Americans' mesocosms were more like giant silos placed on land. The Belgium facility was a small lake and contained a million tons of water, but there was no control facility, so experiments had to be run in sequence.

In the 1980s, we started a program in the People's Republic of China (PRC) at the seaport of Xiamen (Amoy). This program was started as part of Canada's foreign aid, and the International Development Research Centre in Ottawa was responsible for funding it. There were two principal investigators from Canada: Dr. C .S .Wong, who worked with the Department of Fisheries and Oceans, and myself. The Chinese government also appointed two principal investigators. The Chinese were interested in contaminants released during dredging of the port of Xiamen and the mesocosm was an ideal facility for pre-testing. The logistics of starting a mesocosm program in China were horrific, and it was very much to the credit of our field coordinator, Frank Whitney, and to our technical staff, that we ever arrived in China with the equipment needed to make the program work. Customs formalities, alone, for our equipment at the port of entry (Shanghai, Beijing, or Ghanzou), were seemingly designed to assure failure of the program before it could start.

There were other aspects of life in China during the 1980s that we found difficult. While individual scientists were academically excellent and friendly, there was always a counter current of strangeness about our living experience. For example, when we stayed at an hotel once, my wife, Carol, noticed that we had a beautiful carpet in our bedroom.

"But it's disgusting," she said in a loud voice within the privacy of our room. "This carpet has not been swept since it was laid down!"

The same evening, I happened to comment that there was no fan near the bed, which would help in going to sleep in the stifling atmosphere. The next day, on returning to our room, the carpet was spotless and there was a new fan by my bed. A few days later we went up to the roof of the hotel to see the view. On the way up, we passed a floor with what appeared to be an enormous telephone exchange; lots of people listening with earphones and tape recorders.

"So, that's how we got our carpet cleaned and a new fan!" exclaimed my wife.

This eavesdropping continued throughout most of our visit to China. When we traveled up by train to Beijing, our Chinese interpreter said he would leave us in the carriage for a few hours, but would we look after his briefcase while he was gone? He placed it on the carriage floor. During the two hours he was away, we took the opportunity to discuss plans for the Chinese training course that we were going to hold at our university. We noted that on the list of attendees, the interpreter's name was included, to which I objected, since we had our own interpreter for the course, and it would be a waste of an airfare to include the Chinese interpreter. The next day, when we went over the list of attendees with the Chinese officials in Beijing, the interpreter said, "Oh, by the way, I have decided not to come to Canada!"

This kind of snooping was ever present in our relations with the Chinese and it certainly made us feel a little uncomfortable, although, individually, we liked many of their scientists, very much. In particular, Professor Li Guanguo, who was a strong supporter of our pollution experiments. He was a very mild-mannered person, who dressed in western clothes, while most of his colleagues wore Mao suits. He had studied in the USA prior to the revolution in 1949, and had returned to help build a better China. However, because he was not a member of the Communist Party, he was often sidetracked at our meetings, and today it is barely possible to communicate with him in China. It has been an unfortunate error on the part of the PRC to have treated such a loyal citizen in this manner.

The Chinese were generally hospitable when it came to looking after us on the weekends, and in arranging for special banquets and visits to historical sites. Our life with them was not without its amusing side too. While most of the Chinese scientists could speak English, our taxi driver, who took us back from work to the hotel everyday, could not. He drove incredibly slowly, and in the 35°C heat and 90 percent humidity of Xiamen, we naturally wanted to return to the hotel for a shower and a beer, as soon as possible. We learned that he would drive much faster if we mentioned having diarrhoea (a realistically common complaint in a tropical country). So we learned the Chinese word, which was *lào dúzia*; the slightest suggestion of our supposed condition resulted in our driver stepping on the gas all the way home.

Another time, we inquired about laundry facilities in our hotel. There did not seem to be any, so we asked how we were going to wash and dry our clothes. In such circumstances we were always asked to wait, while someone decided on an official answer. Two days later, a pair of long bamboo poles appeared on our balcony and we were told that they were for drying our washing. The washing part seemed to be up to us, and although we were willing to pay someone, this was not acceptable. Nevertheless, the whole of our visit to China was a rewarding experience and gave us a new dimension on oceanographic research.

Just before I left China on my third visit, I picked up a book by Sidney Shapiro, which was written in praise of Mao and the revolution. Sidney Shapiro was an ex-GI who went to China in 1947 and became a Chinese citizen in 1963. He had a blind allegiance to the communist Chinese cause, and so I was interested to read his thoughts on why this rather brutal society was so magnificent. I have a habit of writing notes in the margins of books with which I violently disagree, and by the time I finished Shapiro's book, I had pretty well covered every page with comments. I did not want to keep the book, so I left it in a drawer of the last hotel room where we stayed, in Beijing. No sooner was I in the taxi on the way to the airport, but a hotel clerk came running out with the book. He had a pleased expression.

"Oh doctor, you forgot your book. Very luckily I found it in your room," he said.

"Well, thank you for being so diligent," said I, feeling that this was not the first time that my room had been searched, but probably the last.

I went out to the airport and was careful to leave the book in the taxi while heading for the check-in desk. Again I was pursued, this time by the taxi driver, who would not take a tip for his careful efforts. What was I to do? I did not want this silly book, it was heavy, boring, and I did not want the remarks inside attributed to me. I went into the men's washroom after getting my boarding pass, left it on top of the water closet and rushed out to the customs and immigration gate, never looking back. This time it worked, and I was relieved of both the burden on my thoughts and the extra luggage.

In contrast to our work in the PRC, we also spent some time carrying out research and teaching in Taiwan, at the University of Sun Yat Sen. On the social side, this was a very rewarding experience because the Taiwanese are among the friendliest people in the world. I joined a local tennis club and played practically every morning at 6 a.m., before it became too hot.

Free to Chart an Independent Course

Although there were many members who did not speak a word of English, they always made me very welcome and assured me of a game as soon as I arrived. It would be unfortunate if this hospitable society was one day destroyed by being submerged in the less spontaneous society of the PRC.

Taiwan is full of unexpected signs and expressions that are supposedly acceptable English. This language might be called "Chinenglish." On one young man's sweatshirt, emblazoned with an American flag, we read the motto, "In Dog we Trust." Two street signs, which I photographed in Kaohsiung read: "You Shine Jewelry Co." and "Der-rong Dental Clinic." In the People's Republic of China, we had several living examples of Bud Abbott and Lou Costello's "Who's on first?" The mixture of the English language and the limited number of Chinese names led to a multitude of confused exchanges about Who, Wu, Wong, Wrong, Yu, You, Me and Mi. Archaic English is sometimes learned from old textbooks. I was asked if my wife was a "chatterbox"—an expression invented by Charles Dickens in 1840, and later used by Thomas Hardy, but not common today, at least in North American conversation. One hot day, when asked if I wanted to take "forty tinks" after lunch, I agreed to the suggestion, and while I could assume that "tinks" were winks, I could not explain to my host why forty was such a significant number.

The best use of the English language that I encountered in my travels was in India. Some Indians have a masterful use of English words and pronunciation. For a while after the Second World War, there were a number of Indian nationals hired by the Japanese government to give news commentaries in English. Not being a linguist myself, I constantly admire all attempts to speak our language. While the gentleman in China who insisted on greeting me every morning with "Good night, Dr. Parsons" was a source of amusement, I don't find it insulting to enjoy such exchanges. My own *faux pas* in a foreign language have also been a source of humor to me. One such incident occurred in Chile. I was dining with my friend Renato Quinnones and asked him if I might try my Spanish on the waiter. I said that I would conclude our meal with an opening expression for "I've had enough for today," followed by an appreciation for the food.

I started: "Bastardo por hoy"—I got no further. The waiter looked alarmed, Renato dissolved into laughter and I was confused.

"You just told the waiter he's a bastard," exclaimed Renato. "I will explain to him that you meant to say 'Basto por hoy'."

The Sea's Enthrall: Memoirs of an Oceanographer

While life could have its lighter side, there was serious work to do on the ocean's environment. Aside from our research projects, I felt that it was a role for a university professor to take part in some of the public debates about damage to the environment. I considered myself an environmentalist and had some publications in scientific journals showing effects of pollutants on marine organisms. (One such study was our work on the accumulation of mercury in marine animals on the Fraser River mudflats, which I discuss in a later section.) However, when a particular story in the press seemed to me to be an exaggeration, I challenged it in 'Letters to the Editor.'

One such was the reporting of an outbreak of sylvatic plague in Borneo, due to the World Health Organization's (WHO) program of spraying DDT to combat malaria. The full story was published in the *Journal of the American Institute of Planning* in July, 1971. It was claimed that the pesticide was picked up by cockroaches in the thatched roofs of houses; these were eaten by lizards, who further concentrated the DDT; cats then ate the lizards and died. This, in turn, gave rise to a plague of rats carrying fleas, which caused the sylvatic plague. The story was exposed as being totally inaccurate by a person who had worked in the area of North Borneo, and who gave a paper on the subject at the Environmental Management Symposium, held in Vancouver in October, 1970. Further, even though North Borneo is pretty remote, I was able to contact a WHO official there, and in a letter to the *Vancouver Sun* (December, 1971), I was able to relay, point by point, what actually did happen, the chief point being that there had been no sylvatic plague in the area at the time when the incident was alleged to have occurred, in 1956-58. There were some deaths of cats in the area, which had been attributed to the use of dieldrin (to which cats are highly susceptible), but not DDT. However, a great story had been put together about DDT, and once it got into the press, it was very hard to persuade readers that it did not fit the facts.

I had a similar experience with the building of the Aswan dam in Egypt, which was described as an ecological disaster in the North American press, mainly, I suspected at the time, because the Russians had built it. I visited the area to assess the downstream effects of the dam on the Nile estuary, at the invitation of Dr. Gordon Shrum, who was chairman of BC Hydro at the time. Many of the reports about the dam having caused an ecological disaster were false. More recently, in the early 90s, Egypt was the only country in Northeast Africa that did not suffer the crippling effects of a long drought, thanks to water stored behind the dam, in Lake Nassar.

Free to Chart an Independent Course

Most of the international flights arrive at Cairo airport in the middle of the night. An enormous line-up forms in front of Customs and Immigration with passengers being processed at a geological pace. When I arrived, a senior immigration official walked up and down the long line, occasionally looking at a passport and asking questions. He approached me. "What exactly is the nature of your visit?" he asked.

I said that I was a professor sent out by my government to discuss the effect of the Aswan dam on the ecology of the Nile estuary.

"Ah, a professor," he said with an air of trying to make the two of us jointly impressed by some mutual importance. "You like, I take you through formalities very quickly?"

I had to think fast. Why should I get special treatment? "Oh it's okay," I said trying to hide my impatience. "I will take my turn."

"Oh, very well," said the official. "I come back later."

He left, and came back again in about half an hour, when I had advanced all of five paces.

"You like, you come with me now, professor?" he asked with a big grin.

I had seen him lead off a couple of other passengers and not having heard any wild complaints, I said, "Well maybe that just might be helpful."

We moved through Customs and Immigration with the speed of a Nozomi bullet train, only to come to a crashing stop at the Foreign Exchange counter.

"You want some Egyptian pounds, yes?" says he, grinning from ear to ear.

"Well, yes, I suppose a hundred dollars worth will get me started."

"Oh, too little, too little," he said, "Cairo very expensive and you have not even left the airport, yet." The significance of the subjunctive clause did not escape me. I was not actually free of this weasely individual.

"Well, I will change a hundred for me and fifty for you—how would that be?" I asked.

"Oh, not necessary, not necessary. But very kind, very generous," he exclaimed with a pretence of utter surprise. "Now you are free to go, no more formalities needed. Have a nice day." With that, he disappeared like Rumpelstiltskin, to extract more gold from another passenger in the line-up.

I went outside the airport, only to be immediately surrounded by taxi drivers. "Taxi, suh, taxi suh. Very quick, Cairo," they yelled in chorus.

Seeing a small Mercedes bus marked Airport/Hotels, I fended off the drivers and told them that I was going to take the shuttle. "Oh, very good, suh, have a nice seat inside," one said. The only other occupant was a heavyset man who just said "Da, da" to anything I asked about prices or schedule. After about twenty minutes of waiting with no driver, I got out of the bus and on closer inspection, noticed that it had four flat tires.

"You ready for taxi now, suh?"—a polite way of asking if I had had enough. I told my Russian friend that the bus was not going anywhere, grabbed my bags and asked the price of a taxi.

"Just five pounds," was the answer. We hurtled out of the airport and started across the desert. I was just beginning to enjoy the beautiful star-lit sky and the miles of sand all around me, when the taxi ground to a halt. The driver turned round with a big smile. "Five English pounds, yes?"

That was about three times the agreed price, and this was not really a question, but a demand, given our isolation. We arrived at the Hilton but further difficulties ensued when I refused to get out of the cab, after paying the driver his five English pounds. I wanted to see him take my luggage out of the trunk. This was one dodge that I had been warned about in advance, regarding the tendency for taxi drivers to rapidly disappear after payment, with the traveler's baggage secure in their trunk. The doorman assisted my exit and, at last, I could start to think about why I had come to Egypt in the first place.

Mr. Asad, an oceanographer who had studied in Canada, met me at the Hilton, in Cairo. He did much to guide me through my visit, which took place before the signing of a peace accord with Israel, when the country was still hostile to North America. At one point, to his alarm, I took a photo of the Nile, near my hotel.

"Put the camera away immediately!" he exclaimed. He pointed to a line of barges on the Nile. They belonged to the military. By photographing them I could be mistaken for an Israeli spy. It was only a year later that Egypt was at war with Israel.

The next day, I had to wait for a local travel permit to go to Alexandria. We decided to take a taxi out of Cairo to see the Great Pyramid. The temperature hovered around 40°C. Still dressed in a jacket and tie, I sweltered inside the rickety vehicle. Our throats were dry from the dust of the city streets. We drove past decaying buildings, always trying to avoid the large craters in the road. Around us jostled bikes laden with boxes and buses groaning under the overload of passengers. Children, donkeys, and camels often blocked our path. As we sat, stifling, in the taxi, the odors of sewage, and hot spices from roadside cafés, combined with the cacophony of sounds, to assault our senses.

Free to Chart an Independent Course

Suddenly, relief! The road ended and we were in the desert, gazing up at the 500ft-high Pyramid of Cheops, flanked by its two neighbors. It towered above us, the massive building blocks laid bare by centuries of wind, and sun-aligned to some mystical point defined by architects of a past age. The stones, their former limestone covering long gone, felt hot and rough to the touch, etched by eons of sand erosion. Air shimmered in the heat, and the solid structure became almost a mirage. Imagination ran riot. I felt the power of the construction, invested with the lives of thousands of laborers, given and taken as homage to their god kings.

Coming down to earth, I foolishly agreed to a camel ride around the base. I had not realized how high I would be perched, with no safe means of straddling the animal. As Asad and I started to sway our way towards the pyramid, the heat and motion began to affect me. The ground looked very far away. The camel grunted, ominously, and dribbled thick saliva as it turned its head towards me.

"I want to get down," I yelled to Asad. There was much chattering in Arabic between him and the camel driver. The animal halted and started its slow and awkward kneeling posture. I jumped off its back, with great relief to have my feet once again on the ground.

We arrived back in Cairo tired and dusty. Taxi driver entrepreneurship had delayed us, as drivers haggled over the price. The return journey, it seemed, was priced in proportion to our desire to get back. Capitalism was alive and well, even here, in a sea of sand.

We traveled down to Alexandria the next day, to look at the Nile estuary. A group of Belgian engineers had been hired by the Egyptian government to investigate erosion on the coast. Several coastal hotels had fallen into the sea, and it was claimed by environmentalists that this was due to a lack of sediment from the Nile, after construction of the dam. The engineers had concluded that changes in erosion patterns were largely determined by wind direction. With the shifting of storm centers in the North Atlantic (known as the Azores Low), erosion patterns had changed many times during the history of the country. They did not support the idea that erosion was highly dependent on river flow, which was one of the ecological "disasters" claimed by the environmentalists. Similar beach erosions occur locally in many coastal areas, and sandy summer beaches can turn to rock in winter with a seasonal change in wind direction from Northeast to Southwest.

In the evening, Mr. Asad invited me to have dinner at his house. Mrs.

Asad greeted me in a flowing crimson robe with long sleeves, known as a 'party' galabia. Middle Eastern music played in the background, and we all sat around in a circle, facing a low table. A young female assistant served us drinks, which were either sugarcane juice or Karkadeeh. The latter is a refreshing drink made from hibiscus petals. Mrs. Asad had prepared what was, to me, a very exotic dinner of Middle Eastern dishes. These included falafel; Kishk made from chicken, yoghurt, garlic, and onions; fish Kufta, which was highly spiced deep-fried fish; and Biram Ruz, which was the closest thing to rice pudding that I had experienced outside of the UK. However, the *piece de resistance* was the dessert: Mrs. Asad had proudly saved one packet of Betty Crocker cake mix, which was a hot item in Egypt, where trade with the west had all but ceased. She bore the confection proudly to the table and placed it in front of us with great ceremony.

Our conversation ranged from the problems of the Middle East, to the time that the Asads had spent in Canada. As a dinner guest, I usually make a point of leaving at 10 p.m. However, the pleasant ambience and genial company kept me there much later. It was particularly interesting for me to dine at the Asad's home because it was the only time, in traveling to many Moslem countries, that I was invited as a guest to a Moslem home.

Returning to other anthropogenic issues of global importance, there was increasing concern at the number of oil spills that had occurred in the oceans. Tankers had become larger and larger between the 1950s and 1980s, and when a 400,000-ton tanker, such as *Amoco Cadiz*, went on the rocks, there was a very significant impact for at least 100km, or more, along the coastline. Birds and benthic (ocean floor) animals were the most severely impacted, with mass mortalities. Seaweeds and marsh grasses were severely damaged, but fish were not greatly affected, since they could mostly leave the polluted areas. The press made the most of such oil spills, since they are a very photogenic form of pollution, with so many mass mortalities.

Some oil spills were much more harmful because they impacted certain specific fisheries. One such, was a relatively small (less than 40,000 tons) spill that occurred in the Sea of Japan. This is an area in which there are many aquaculture farms, and so the plants and animals could not be moved, and were thus killed. Other oil spills occurred in the open ocean and broke up at sea without causing damage to the shoreline, but they still affected thousands of sea birds. A few oil spills had a much greater impact than their size implied. The spill in Valdez, Alaska was relatively small (less than

30,000 tons), but the Alaskan coastal current carried the oil over about a 1,000km of coast line, causing widespread damage. The largest oil spill occurred during the Gulf War, when Saddam Hussein of Iraq released over a million tons of oil into the sea. In my opinion, research on oil spills is not very profitable, compared with other forms of pollution. It is better to prevent the spill in the first place, by having proper navigational aids, avoiding human error and malice when possible, and mandating the use of double-hulled tankers.

It is difficult to produce photographic evidence of more insidious pollutants, such as tributyl tin, an antifouling agent which is so highly efficient that at very low concentrations it kills the larval forms of most marine organisms. Fortunately, it has been banned from many marine uses, but may still be used for warships and very large commercial vessels.

One of the most controversial forms of marine pollution to arise recently has been the effects of aquaculture on the marine habitat. In some coastal areas of Canada, Norway, Scotland, and Chile, for example, large numbers of salmon are held in open net pens in coastal embayments. The waste from these animals, which includes uneaten food and feces, is allowed to settle on the bottom under the pen, where it can produce an anoxic (low oxygen) zone, at depth, over a small area. The environmental attack on salmon farming goes much further, however, and deliberate attempts are made to thwart the industry, both on site and in the sale of farmed fish. There is no doubt that as a very new industry, new regulations are required, but the environmental movement wants a complete ban on open net farms. The location of these farms in 'on-shore' tanks would be too costly for the farmer, but that is what is demanded.

It is necessary to go behind the scenes a little, to find out why this obvious solution to a shortage of fish has come under such attack from environmentalists. In fact, the main attack has come from the fishermen who capture wild stocks of salmon, and who have generated many of the reasons for shutting down fish farms that are now reiterated by environmentalists. The reason for the strong opposition of the fishermen appears to be more economic than environmental, although this is never stated. As the fishermen have exploited fish stocks world wide, to the point of severely decimating their number, the price of fish has risen, so that even catching less fish does not matter, if the unit price continues to rise with the scarcity of fish. However, enter aquaculture, and you have a price stabilization which lowers the soaring cost of fish. This is not good news for fishermen exploiting wild stocks. Environmental groups have tended to side with the fishermen. The over-exploitation of wild stocks has largely gone unnoticed. In the extreme case, several restaurants have sided with the fishermen in refusing to serve farmed salmon, in spite of the damage such a policy inflicts on wild stocks.

Some of the extremes to which the environmental movement has gone, in order to target fish farms, are worth noting. On a large van selling wild-caught salmon, a slogan was carefully painted, "Our Fish Do Not Do Drugs." The inference here was that farmed salmon had to be inoculated against disease, which was not the case for wild salmon. However, such inoculations have to be approved by Food Inspection Agencies. In Canada, a 1999 report by our agency listed 47 medicating ingredients that were approved for pork, 41 for chicken, and so on, down to salmon, for which only one medicating ingredient was approved. The van's display, of course, failed to point out the widespread use of medication in raising other agricultural products.

In another instance, it was claimed that since farmed salmon were fed fish pellets, this was a wasteful use of fish resources in the ocean. The use of fish pellets as an animal feed comes mostly from Peru and Chile, where very large amounts of anchovy and mackerel are harvested, for which there is no market, other than as dried animal food. It was this harvest of millions of tons of very small fish that gave rise, in the 1960s and 70s, to fried chicken outlets that became so popular as to compete with the hamburger stands. However, chickens eating fish pellets have a maximum conversion efficiency of food to flesh of about 10 percent. Fish living in the sea do not need large bones, or energy to fight gravity. They have up to 30 percent efficiency in converting high quality protein to fish flesh. It makes sense, therefore, to use this fish food for farmed fish, rather than chickens.

Today, whenever there is a decimation of some salmon run, aquaculture is blamed as having interfered with the ecosystem. Currently, the focus is on sealice, which breed more easily on farmed fish than on fish in the wild. Whether this particular example of an impact is real or not is still not known. What is clear, however, is that long before there were fish farms, there were very large fluctuations in salmon stocks from year to year, and at least some these might be attributed to over-fishing, rather than to the impact of fish farming. There is no doubt that fish farming will require regulation of various kinds, but the environmental attacks launched by fishermen and such organizations as the Suzuki Foundation, are hardly justified, in my estimation.

Free to Chart an Independent Course

3. Sailing against the current

There are times in life when we try to convince people as to the correctness of a certain idea, only to find that it is systematically rejected. It is not that the idea is necessarily wrong (although it may be), but that there exists a persistent wall of opposition. This was the case in my efforts to introduce a more basic scientific approach to the study of fisheries, through an understanding of fish ecology. Although this opposition is something that occurred at my university, the problem of why fisheries have been so badly managed in the world's ocean is widespread; my experience is only one example of a problem that has been repeated worldwide, at many institutions.

We live in a watery world. Water is the main constituent of our bodies, and water occupies seven-tenths of the world's surface. By volume, the landmass of our world hardly makes a splash in the oceans that are ten times bigger. Choosing to study oceanography was coincidental with my boyhood love of the sea. The creation and survival of life on our planet became a subject I found vitally interesting.

The oceans are the cradle of life and the cooling system for the planet. In them, volcanic emissions have deposited their waste for millions of years. More recently, humanity has done the same. The oceans are our highways between distant countries. The oceans are a benevolent source of food which, on average, makes up at least 20 percent of the world's supply of high quality protein, or more than 80 percent, in some maritime countries.

Unfortunately, national boundaries did not extend very far into the oceans, until the 200-mile economic zones were established in the 1960s. Thus, the exploitation of the oceans' protein resources was a "free-for-all," with a "winner-take-all" policy prevailing, up until the second half of the 20th century. This has contributed to the over-exploitation of fish stocks. The huge investments made, post-war, in the fishing fleets of such countries as Korea, Spain, Taiwan, Poland, Russia, and Japan had to be paid for by a cash return of fish. In the absence of an understanding of the ecology of these fish, a science was developed for management through regulation, but without understanding. This allowed for over-exploitation of both national and international fisheries. A new approach to the management of fish stocks was needed.

The scientific management of the oceans falls under two headings:

Oceanography and Fisheries. These two sciences have had separate developments, and are financed independently in practically every country. Fisheries science has, historically, depended heavily on data from commercial catches of fish. This has proven to be an unreliable way to manage fisheries. I have felt for a long time, as have others, that the whole basis for fisheries science was flawed. What was wrong was the lack of recognition for the information being generated by a different group of scientists, the oceanographers. This information could explain how the oceans were ever-changing, as an environment for fish. These changes could account for the missing information in fisheries science.

To put my life as a scientist in context, it was particularly through my work on fisheries that I felt a great need to help to develop a new oceanographic discipline, which combined the needs of fisheries management with the science of oceanography. The science of oceanography covers at least four disciplines. There are the physical oceanographers, who study the movement of water masses and the heating and cooling of the ocean waters. There are the geological oceanographers, who are concerned with bottom topography, its composition, and the giant cracks that run around our globe under the ocean surface. The chemical oceanographers are concerned with the composition of seawater, the amount of nutrients present, the occurrence of pollutants, and how some of the elements and compounds have changed over time.

The biological oceanographer's role is much more difficult to define than the previous three. Originally, biological oceanographers studied those plants and animals which move only with the ocean currents (i.e. *plankton*, meaning "to drift"). The other half of the biology of the oceans, the *nekton* (i.e. meaning "to swim"), was left largely to the fish biologists, of which there were two types. Some fish biologists studied the physiology and metabolism of the different fish species, and amassed much important data on these functions. Other fish biologists were concerned with the management of fisheries, and these persons turned largely to theories on population biology and data on fish catch, to describe the superabundant quantities of some species of fish in the sea, and the effect of fishing on these stocks. It was this science that failed, and it was here that I felt a new discipline was needed.

The main concern of the fisheries scientists was the exploitation and preservation of these stocks. Their theories were varied, but they contained a central theme, i.e., the number of young fish that entered a fishery was proportional to the number of adults present to lay eggs. The theory is generally referred to as the stock/recruitment theory. The principal features of this theory and the historical damage that it has inflicted are as follows:

⋏ It has generally relied on the use of commercial catch data to show how many fish were present in the first place, how many were caught, and by difference, how many survived to create a new generation of fish. These data are known to be inaccurate. The subject is further complicated by different species of fish having different life cycles, but at one time or another the theory has been applied to almost every commercial species of fish and has consistently failed.

⋏ As a result of this failure to manage the fish stocks of the world, we see, today, declines in certain well-known fish resources by 50 percent or more, such as the cod off Newfoundland, compared with historical data on the size of these stocks.

⋏ Furthermore, large fish in the sea are being replaced by small fish. This has happened in the Black Sea, the North Sea, and the Western Atlantic. Fisheries for mackerel and herring have been replaced by fisheries for the much smaller species, sprat, and sandlance.

⋏ In other places, shark-like fishes, known as Elasmobranchs, were replacing the edible fishes, such as cod. In Canada, The Harris Report, written at the time of the cod fishery collapse off Newfoundland, recommended stopping fishing so that the cod would come back. They did not. There was more ecological disturbance than the fisheries experts had realized.

⋏ In general, the fisheries of the world are in very bad shape, and little is being done about it. Our local newspapers picked up this dilemma, with such titles as "Herring Season in a Pickle"; "Net Result is a Mystery"; "Great Cycles Baffle Fisheries Researchers"; "Salmon Population Surge Mystifying," etc. If these headlines belonged to some other branch of science, such as medicine or agriculture, we would never trust heart operations or our food supply again. Yet, because of the remoteness of the oceans, we have somehow come to tolerate such reports.

I believed that the main cause of the discrepancy between fisheries science and our unmanaged fisheries was that the science had failed. The reliance upon commercial catch data, as the mainstay of fisheries science, was not justified. The populations of fish in the sea vary with changes in ocean climate, which means that it is the ocean environment that determines the abundance of fish stocks. It is this same ocean environment which can change and, through the ecosystem of the sea, lead to less abundance. This environmental "wild card" was never included during fifty years of theorizing on the management of fisheries. Some scientists argued that they could not achieve understanding of the ocean environment through ecosystem modeling of the food web and climate change. It was said to be too complicated in the foreseeable future because of the vastness and complexity of the oceans. Yet, applying national and international regulations to fisheries, without understanding the basic ecology of fish in the oceans, is a flawed use of science.

For nearly a century, biological oceanographers had been studying the plankton of the sea. It had become clear that its existence was tied to changes in the climate of the oceans. More plankton was produced if nutrients were brought to the surface by strong wind action. Different kinds of plankton were produced, depending on different water masses and, in general, the fish of the sea followed the plankton because that is what many of them ate. Thus, if our team studied the plankton, we might understand more about the fluctuations in fish populations. We had already demonstrated in our Great Central Lake experiment that enhanced plankton production could greatly increase the production of fish.

It was clear to us that fisheries science needed a new approach to biological studies of the ocean. As with the evolution of medicine, ocean biology had to evolve. Medicine, in the 1930s, was a perfunctory science that removed tonsils from a whole generation of children, prescribed tonics that had multiple ingredients of largely unproven effect, and there was an almost religious concern with daily visits to the toilet. All of this changed when other branches of science, including physics (e.g. x-rays), pharmacology (e.g. antibiotics), genetics (e.g. birth defects), and biochemistry (e.g. vitamin deficiency) invaded the science of medicine and made available new techniques and new treatments. Today, we have an acceptable science of medicine, which is not an exact science, but a vast improvement over that of the 1930s. The same process had to be introduced in fisheries science, if we were going to manage our ocean resources. We needed to understand the physics, chemistry, and biology of the oceans in such a way that we could come up with some indices of prediction for the management of our fish resources. This was where biological oceanography could be combined with fisheries science, in what we might call "fisheries oceanography." This we could define as an holistic approach to the study of ocean ecosystems, for the benefit of fisheries resources.

My work linking plankton to the fish of the sea took several directions, in addition to the Great Central Lake experiment. In 1985, I wrote an article about Pacific fish stocks for the Organization for Economic Development (OECD), showing that many changes in fish populations were caused by changes in ocean climate and not necessarily by over-fishing. In fact, it was probably a combination of these two effects that caused the dramatic changes listed above. However, we really need to know more about how ocean ecosystems work, in order to understand how to manage fisheries. To this end, I can give a short description of some of our own work which, together with many additional studies by other oceanographers, could form the basis of a new management strategy.

The example from our research was to develop some ideas on food chain structure in the sea. Other scientists had noted that photosynthesis in the sea was sometimes due to very small phytoplankton, called flagellates (10 microns, or less in diameter), and sometimes due to very large phytoplankton, called diatoms (often greater than 100 microns in diameter). In 1977, a visiting scientist from Hamburg, Dr. Wulff Greve, and I, evolved an idea based on our experiments that the flagellate food chain tended to lead to jellyfish, and the diatom food chain led to fish. In 1979, I had some ideas to add to this concept during my sabbatical leave in Australia, and these came from the fossil record. In brief, this record showed that jellyfish were around in the post-Cambrian, 500 million years ago, but that the Age of Fishes was only about 300 million years ago, while whales evolved in the oceans 100 million years ago.

The importance of this to our studies was that whales require 200 times more energy, per unit weight, than jellyfish, and fish require seven times more. Also, the intelligence of the organisms in this hierarchy increased with their energy consumption. To me, that meant that something happened over time to the primary producers of the sea, to make them more productive agents in the evolution processes of ocean ecology. I ascribed this increase in productivity to the larger phytoplankton since, in general, bigger prey items require less energy to consume and provide more energy per bite than very small prey items. The size of phytoplankton cells appears also to have increased over time, the large diatom cells being absent until the last 100 million years, which coincides with the evolution of whales. It also appears that the intermediary zooplankton that consumed the phytoplankton were quite small, Post-Cambrian, but later, much larger zooplankton appeared.

An analogy exists in that primitive humans could not evolve an intelligent society, until their food supply was assured through agriculture, and a mechanized humanity could not evolve further, until its power supply of about 60 barrels of oil per capita, per annum, was assured. Space-faring humans requires an even higher per capita energy supply. Other corollaries of this hypothesis are that longevity increases and fecundity decreases, in conjunction with higher energy consumption. For example, jellyfish are more prolific than fish, and fish are more prolific than whales. In addition, since the jellyfish require much less metabolic energy than fish or whales, they have the potential of creating a far greater biomass in the ocean than either of the more metabolically expensive animals. These ideas of patterns in Nature are called Macroecology. Not every animal or plant has to fall into this pattern. Even the greatest macroecologist, Darwin, when providing data in support of his theory of evolution, resorted to such terms as "in general" and "sometimes." So it is that my hypothesis applied to food chains of the sea is controversial, but it has been supported by some recent events and observations.

In the Black Sea, in the 1990s, Professor Zaitsev reported large numbers of jellyfish had started to take over from previously abundant fish stocks. In fact, over 10 billion tons of jellyfish came to dominate the entire sea, under conditions that might have been attributable to either over-fishing or pollution. Since the large diatom cells are more susceptible to pollution, their replacement would have been by flagellates favoring a jellyfish food chain; the removal of fish by fisheries would also favor a jellyfish ecology. The same thing happened in the Adriatic, and then in the Bering Sea. In the latter, there was very little in the way of pollution, but it was where one of the biggest fisheries in the world was located. Large numbers of jellyfish started to appear. These changes in jellyfish numbers would fit our ideas, but they could also have been caused by changes in climate. This controversial hypothesis opened up some very new considerations for fisheries management, based on the food chain of the sea, rather than on the older theory of population dynamics.

Another place where our hypothesis fitted the facts was in the upwelling regions of the world, the Benguela Current, the Canary Current, the California Current, and the Peru Current. These areas are all dominated by diatom ecology, and they are also rich in fish and whales. By contrast, Cnidarean ecology (Cnidarians include the jellyfish and the corals) dominates in convergent seas, such as the west coast of Australia, the Caribbean, the coast of Madagascar, and parts of Southeast Asia. These seas are also dominated by flagellate ecology and not by diatoms. The jury is still out on these ideas, but it is essential that biological oceanography includes, in the future, all of the major predators of the sea. There always seems to be research money available to study commercial fish, such as salmon or cod, yet jellyfish may be posing the biggest threat to world fisheries, today. The case should be made for pointing research funds toward the understanding of jellyfish ecology.

In yet another approach to the problem of linking the very small phytoplankton to the fish of the sea, Dr. Sheldon, who came as a post-doctoral fellow from the United Kingdom to study with me, proposed a size-related hypothesis of marine food chains. He said it was relatively unimportant to know what species were present at any level, but it was important to know the size of the organisms present. We published a paper about this size spectrum of organisms in the ocean, in which we viewed the ocean as a sort of continuous compacting machine that pushed small things into larger things, and so on *ad infinitum,* at least, up to the size of the largest fish. It sounds a very crude approach to be scientific but, in fact, it works in some ways, and one thing it tells us is the "carrying capacity" of the oceans. This idea was developed by Dr. Sheldon, after he moved to the Bedford Institute in Halifax, Nova Scotia. It is an important concept, because it defines how much fish can be taken out from any ocean area. I believe that Dr. Sheldon's approach will be very useful to fisheries managers in the future, and should have been utilized earlier.

An opportunity to work on Sheldon's idea came when I was an invited researcher/lecturer at the University of Sun Yat Sen, in Kaohsiung, Taiwan. I collaborated with Professor Yu-ling Lee Chen in trying to estimate the production of different sized fish in the coastal waters of the South China

Sea, from a knowledge of the amount of photosynthesis in the same waters. Measuring the amount of photosynthesis was not difficult, and the total fish catch in the area was well reported. However, the size spectrum of fish needed in the calculation was not readily available. To get the size of the fish caught, Professor Chen and I went down to the local fish market and bought random samples of all the fish being sold. We were a high nuisance factor to the local vendors! The samples were taken back to the laboratory and sized. From the calculations that followed, we were able to compare the efficiency of fish production in the tropics with that of temperate waters in the Northwest Pacific. While the latter produced more fish, the actual efficiency of production was higher in the tropics than further north.

My collaboration with Dr. Sheldon was also a factor in another event. We used a piece of medical equipment known as a Coulter Counter©, which had originally been designed to count red blood cells. We used it to count and size phytoplankton in sea water, and then measured how much was eaten by the zooplankton. Measuring phytoplankton in sea water is not so very different from measuring red blood cells in blood. As it turned out, the equipment was well suited for counting phytoplankton; it was also an example of how important it is to develop a new technique for innovative ventures in science.

I published a paper on my findings in the *Journal of the Oceanographic Society of Japan*, and it became a citation classic. This was the first, and at that date, the only time that the society journal had had one of its papers reach this level. I believe that it was mainly because of this paper that in 1988, I became the first non-Japanese scientist to be awarded The Oceanographic Society of Japan Prize for excellence in research. In retrospect, I believe this was one of the first steps that led to The Japan Prize itself, and as such, marked an important point in my career. Prior to this, I had received various scientific awards and recognition in my own country, and had been elected to the Royal Society of Canada, but this was the first time that I had been recognized abroad. I decided to go to Tokyo to receive the award, which was given at the annual meeting of the Society. Subsequently, three of my research colleagues in Japan who came to Canada to study with me, Professors Seki, Maita, and Takahashi, have all received The Oceanographic Society of Japan Prize for their work, either in chemistry or microbiology.

Not all my efforts in trying to establish fisheries oceanography were scientific; some were more administrative. For example, I decided that we needed a Chair of Fisheries Oceanography at my university. Such a position would be a focal point for the science and would bring in additional funds. The subject was becoming popular at other locations, such as the University of Alaska in Fairbanks, and Memorial University in Newfoundland, which had started to canvass the public for similar positions. In both cases, they were eventually successful through donations from the fishing companies, among other sources of income. In British Columbia, I was able to obtain a large donation from a very wealthy prospector who liked eating fish. This anonymous benefactor once arrived at our house and ordered a catering company to bring half a dozen Atlantic lobsters with all the trimmings, so that we could have a feast.

A subsequent donation for the chair came from the Packard Foundation, in the United States. Together with funds from the federal government, the establishment of a chair was successful, although someone changed the name to the Chair of the Ocean Environment and Its Living Resources. I did not think this was nearly as suitable as bringing in the word "fisheries," but our campus had a very peculiar administrative organization when it came to fisheries science. It was here that I was to fail totally in trying to make any impression on the administration of ocean science at my university.

An Institute of Fisheries had been established at the university in 1952, as a joint effort between the university and some employees of the provincial government. At that time, it was mainly the work of the late Dr. Peter Larkin. He was a fisheries biologist who, in 1966, became Director of the Biological Station in Nanaimo, where I was working. We had some excellent directors at the Biological Station, including Dr. Alfred Needler, Mr. Radway Allen, and the internationally acclaimed Dr. Bill Ricker, the last being an acting director for a number of years. All of these persons supported research on the ecosystem approach to fisheries management.

However, Dr. Larkin was primarily a theoretical fisheries scientist, and in the thirty years that I knew him, I was not aware of his participation in experimental or field observations. He was not a supporter of the need to integrate fisheries and oceanography, and I feel that this was detrimental to the cohesive development of ocean science. He was one of a number of fisheries scientists who wrote prolifically about fisheries, while comfortably seated behind their desks. In contrast, it is my belief that all ecologists should spend several months of each year in the field, studying Nature. Dr. Larkin was a genial character who was, technically, a good administrator. He was well known locally, but did not have much of an international reputation. He would often agree with what you were saying. His writings are full of all the right buzz words on ecology and fisheries, mostly borrowed from many, more original, scientists. His writings and apparent agreement, however, could never be taken as an endorsement of one's ideas. He held to his own rigid opinions, and he represented the views of a typical fisheries scientist of the 1950s era.

I had encountered their opinions before, when attending a meeting, in 1962, of the Advisory Committee on Marine Resources Research (ACMRR), in Rome, under the chairmanship of Mr. Sidney Holt, a prominent fisheries biologist. At one meeting, Mr. Holt claimed that the new anchovy fishery off the coast of Peru, which was just starting in the 1960s, would be managed, in detail, by careful control of the adult stock. It was a chance for the fisheries biologists to get it right with a brand new fishery! Professor Uda was present at the same meeting, and he disagreed with this opinion, favoring, instead, greater input from oceanographers. In less than twenty years after this meeting, this fishery became exhausted, but it recovered many years later, with a change in oceanographic conditions.

I am by no means sure that my criticisms of these fisheries scientists would be shared by other scientists. It is simply my personal view that fisheries science *has* to be based on a fundamental understanding of how the ecology of the sea functions, and that there is no short cut solution through population dynamics. Obviously, there are others who do not share this belief, and only time will tell who is right. From my discussions with Dr. Larkin, it appeared that he held views similar to Mr. Holt. Thus, Dr. Larkin's considerable administrative influence, while in the employ of the government, and later, in academic affairs after he returned to the university (where he became Dean of Graduate Studies and later, Vice President), did not help our cause. The Fisheries Institute at the university that he founded was, in fact, discontinued as such in 1968, and renamed the Institute of Animal Resource Ecology. It was significant to me that the new institute only concerned itself with 'animal' resource ecology, as if the plants were not important (was there not something to learn from agricultural science?)

After about a twenty-year hiatus, a Fisheries Institute was re-established on our campus, and I would have liked to see oceanography integrated in this new structure, since fisheries scientists and oceanographers study the same environment. My suggestions fell on deaf ears, as far as the Dean was concerned. Protecting local empires on campus was often more important than advancing knowledge. As Henry Kissinger once said, "University politics made me long for the simplicity of the Middle East."

In addition to the lack of support on campus, not a single fishing company on the west coast donated to the Chair of Fisheries Oceanography. This was not the case in other parts of North America, but then Dr. Larkin was also on the board of directors of one of the largest fishing companies. There was a donation given by a fishing company to the university, but it went to a group of engineers. They planned to research a better way to can fish. The fishing industry has already been totally over-engineered, with respect to the catching and preserving of fish, but here was another typical example of a lack of interest in the ecology of fish, which would protect the resource, while piling even more money into the profit margin of selling the product.

A new Institute of Fisheries was finally set up without oceanographers, who were soon absorbed into geology and geoscience, during a campus drive towards larger administrative units. What was the reason behind the lack of a cohesive approach towards fisheries science on our campus and elsewhere in the world?

Government funding and administration have always separated fisheries research and ocean research. This has been a long-standing separation worldwide, largely because there are a lot of economic and political problems associated with fisheries, while most of the rest of ocean science is concerned with understanding the basic mechanisms of ocean circulation, chemistry, and biology. This has often resulted in fisheries research being more of a probability-based 'quick fix' science, than a science devoted to understanding the fundamentals of ecology. Fisheries oceanography is a science that could be very helpful in bridging this gap and managing fisheries, but it could also disrupt the role of fishing companies in their exploitation of ocean resources. The science requires one to stand back and ask, "Hey, what are you doing to the ocean ecosystem?" This is a question fishing companies do not want to hear, just as logging companies do not want controls on logging. However, the latter have become accepted practice, while the fishing companies are still largely irresponsible in their attitude to ecosystem understanding and preservation.

Again, the other problem is that many of the senior fisheries scientists have given up going out to sea to observe and collect their own data. Their chief source of data has traditionally been the commercial fish catch, and their primary research tool has been analyses of mathematical models on the computer. Real fish do not swim and grow inside computers, where there is little attempt to provide an adequate environmental setting. I have heard the comment more than once that fishermen know more about the fish of the sea than the fisheries scientists! It is inconceivable to me that one can study Nature without observing it. Darwin's theories would never have evolved without his first-hand experience of Nature. In a more recent context, Professor E. O. Wilson's (Harvard University) work on sociobiology did not come from sitting behind a desk, but from his observations of the small world of ants. The great naturalist explorer, William Beebe (1877-1962), probably said it best: "Discovery is seeing what every one else has seen but thinking what no one else has thought."

In oceanography, data collection is a most important role for the scientist. At the University of Washington, a professor of oceanography recently took his whole class of physical oceanographers to sea, so that they could understand the problems of "on site" observations. I have never heard of a fisheries scientist doing this. In fact, I viewed with some dismay the detachment of many senior fisheries scientists from field observations. It often resulted in conflict of interest. If, instead of carrying out field observations on fish, one takes on the role of a director in a fishing company, then the profit motive must dictate that the only acceptable advice to give is how to catch more fish. The problem reminded me of the admiral's ditty from *HMS Pinafore*. I could not resist adapting it. Had Gilbert been an ocean scientist, I think that he might have been sympathetic:

> *"Stick close to your desks, and never go to sea,*
> *And you all may be rulers ... of a fisheree!"*

In the mid 1990s, however, I was encouraged by a trend which suggested that others were beginning to recognize the need for change. In a surprise move, a new faculty member in the Fisheries Institute, Dr. Daniel Pauly, started an ecosystem approach to fisheries research as a "new" science, although there was still no formal association with the oceanographers. It was, at least, a partial move in the right direction to begin to observe the ecosystem. In my opinion, however, it may not be sufficient to be effective, if it is directed more towards another 'quick fix' policy for fisheries, rather than an understanding of ocean ecology grounded in basic research using new data. To me, it is a perverted view of science to study the ecology of any animal without studying its environment.

However, some measure of just how far things *had* changed at this time is given in two quotations by persons in the Fisheries Institute at my university. In 1992, two fisheries scientists Hilborn and Walters, wrote: "We believe that the food web modeling approach is hopeless as an aid to formulating management advice; the number of parameters and assumptions required is enormous" (*Quantitative Stock Assessment: Choice, Dynamics and Uncertainty*, Chapman and Hall, NY, p.448). In 2000, one of the same scientists, Dr. Walters, wrote: "We simply cannot continue to ignore trophic interactions in the formulation of models for exploited marine fishes, particularly in evaluation of policies like protected areas that are almost certain to have a variety of whole-ecosystem effects" (*Marine Ecology, Progress Series,* 208: p.312). The latter quotation comes a long way towards adopting oceanography as a basis for fisheries management, although it fails to fully recognize this need.

These differences of policy between fisheries and oceanography were not unique to our university. They have occurred at many research centers throughout the world during the last five decades. One possible remedy to this situation is to separate fisheries management from science, and locate it in either departments of Commerce or Economics in the university structure. Fisheries ecology could then be pursued by academics interested in fundamental studies on the relation between fish and their ocean environment. In particular, fisheries ecologists should be concerned with the question of just how much fish can be taken from the oceans without upsetting the ocean ecosystem. As an example of this problem, it is now clear that the cod stocks on the Grand Banks have not returned after a twelve-year moratorium on fishing. The advice of fisheries scientists in the *Harris Report* suggesting the moratorium, as a solution to restoring the cod stocks, was clearly misleading because it lacked real understanding of the Grand Banks ecosystem.

There is an analogy in this suggestion to separate the basic science from management. Both insurance companies and climatologists are interested in the weather. Insurance companies pay out a lot of money, depending on the frequency and severity of storms. They use statistical patterns, and possibly some game theory, in order to manage their business. Climatologists are also interested in the weather but they want to understand the fundamental nature of storms, how they form, how they spread, and so on. In this analogy, fisheries management should be based on economics and the application of new theories. Fisheries Science should probably not exist as a subject because it confuses two subjects—commerce and ecology. Fisheries Ecology could be advanced as a science dedicated to understanding the relationship between fish and their environment. At present, Fisheries Science has led to the mismanagement of the world's fisheries, to the point that fish stocks, which make up about 20 percent of our high quality protein, have been decimated. As with the climatologist analogy, study of Fisheries Ecology, if actively pursued as a pure science, could eventually lead into new management strategies for the future.

It would not be fair to assume that there was a total absence of open ocean studies by fisheries scientists. An example of a marked exception, in my experience, was the work of a fisheries biologist, Dr. David Welch. He collected his own data on the distribution of young salmon in the Gulf of Alaska, using a Russian research vessel; the stormy waters and living accommodation for such work add to my admiration of this scientist's originality. Quite recently, I was also very impressed by the work of a young woman who studied, briefly, in my laboratory, but then went on to Cambridge, where she obtained her doctorate. Her name is Amanda Vincent, and she has made a great reputation for herself in the study of sea horses. This small fish was threatened in tropical waters because dried specimens are readily sold as novelties and as medicine throughout the world. This has resulted in their decimation in many areas—a process no different from the over-exploitation of commercial fish. Unlike fisheries scientists in general, however, Amanda carried out basic studies on the ecology of the species, rather than relying on a population dynamic theory. She was so successful as to be able to help the people of Southeast Asia raise their own sea horses, and her work has been reported in the National Geographic and on prestigious television shows. She is a rare example of a fisheries scientist who solved an ecological problem by a "hands on" collection of original data.

Taking another tack, in my position at the university, I felt that I could make a concrete contribution to furthering basic research on fisheries, through the selective publication of the work of other scientists holding views that were similar to mine. I decided to start a new journal on the subject. Publishing houses did not beat my door down to start yet another journal, but I was lucky enough to come in contact with Professor Sugimoto of the Ocean Research Institute, Tokyo. He suggested that the Japanese Society of Fisheries Oceanography (started by Professor Uda) should sponsor an international journal, providing I would serve as Editor in Chief. We discussed this idea, in Tokyo, with other members of the Society and there seemed to be universal support to adopt the suggestion. I was more than happy to be the first editor, and we contacted an excellent publishing house, Blackwell Science in Oxford, England, and they were prepared to undertake the publication. The first edition of *Fisheries Oceanography* came out in 1992, with a cover I had designed. The cover was bright aquamarine, with stylized fishes swimming on a wavy background, which could be interpreted as graphed data or rolling waves. I had invited some senior persons in fisheries and oceanography to contribute to the first number, and this gave it some importance.

I stayed as Editor for five years, during which time the journal grew in size and distribution. I was followed by a senior and very capable scientist at Scripps Institution of Oceanography, the late Professor Mike Mullin. Dr Dave Checkley of the same institution then carried on the editorship, with excellent results. An indication of their combined success with this journal is given in an excerpt from the following letter to Dr. Checkley, from the leading organization for the evaluation of scientific journals:

20.04.03

Dear Dr. Checkley:

I am writing to you concerning a recent analysis of total citation increases over a given bimonthly period performed by ISI Essential Science Indicators (ESI). The results of our analysis indicate that Fisheries Oceanography had the highest percentage increase in total citations in the field of Plant and Animal Science. We would like to include a feature on Fisheries Oceanography in this field, to accompany the results of our analysis....

Jennifer. L. Minnick,
Editorial Coordinator
ISI Essential Science Index

It would seem from this correspondence that both the messenger and the message were finally finding a place in ocean science.

I had started out with high hopes of establishing my university as a leader in fisheries oceanography, a new science for the management of fisheries, but I had reached the end of my tether. Like my butterfly collection in school, with its misspelled species, like my attempt to transfer oceanography to a smaller university, my efforts to amalgamate ocean sciences at my university had come to naught. It was time to "call it quits." I decided, then, to retire early. For nearly thirty years, one man had kept reappearing in my professional career, always one or two positions higher in the administrative hierarchy, and always opposed to the integration of the ocean, fisheries, and atmospheric sciences. He knew how to operate within the bureaucracy of academia and I did not. Now, I felt that I had been shut out again. I might deserve something for effort, but this was, in the end, a "no win" situation for me. However, as later events began to unfold, it proved to be a situation which the Japanese recognize as *sonsite-tokutore*, or, "gain by losing." The ideas first suggested to Dr. Strickland and me in 1958, by Dr. Needler, that we should look for scientific alternatives to population dynamics for the management of fisheries, were now, in fact, beginning to take hold.

While I failed in the organization of this task, there was some progress in the spread of interest in this subject outside my own university, through scientific publications and the establishment of a new journal. The changing of the name of the Chair of Fisheries Oceanography, to which title people had contributed money, was a disappointment to me, as well as possibly being illegal. It was a choice made by the administration to call it the Chair of the Ocean Environment and Its Living Resources, and although I had initiated the need for this position, I was never consulted on the name change. Under the new name it fell short of its purpose, and could eventually become a position devoted to studies on oil pollution.

In another respect, I received more than a passing grade. In 1993, the American Society of Limnology and Oceanography awarded me their Hutchinson Medal. Evelyn Hutchinson had been a professor at Yale University for many years, and was responsible for starting some new ideas in aquatic biology. Among them was his belief in an holistic approach to Nature, in that all parts of natural systems are seen as being interconnected. The award was given to me "in recognition for a career spanning many aspects of oceanography" and for "trying to make ecology predictable by finding and accurately measuring parameters." I owe a debt of gratitude to the anonymous people who submitted my name for this medal.

4. How to train a new crew

It is a very important part of university life to teach and train a new generation, in my case, of oceanographers. I started a number of successful courses at the university, both at the graduate and undergraduate level. My lectures on oceanography were given to classes which varied in size from a large undergraduate class of about seventy students, to the much smaller graduate classes of ten to fifteen students. The latter were a dedicated group, who had already decided to join the ranks of the professional researchers and were working toward higher degrees. All students could evaluate their professors at the end of the year and these evaluations were computerized and compared with other professors'. I would have liked to have been an outstanding lecturer but never was. My evaluation by students was generally above average but never in the excellent category. I was enthusiastic about my subject but unable to put it across in the masterful way of some of our teaching staff. This was a disappointment because, in my own career, I had valued a number of good teachers very highly.

The comments students wrote on the backs of their evaluation computer cards were sometimes personal: "He wears the same shirt for a week"—I have more than one shirt of the same pattern; "I would like you to speak a little slower in your lectures"—helpful; "This is the worst course that I have ever taken"; and "This is the best course I have ever taken." Somewhere in between these comments lay an element of truth, and in general, constructive comments were most welcome.

The training of graduate students differs from that of undergraduates. Graduate students are required to attend some courses, but their work is more dependent on the results of independent research, recorded and evaluated in a thesis. Problems sometimes arise because of the collaborative nature of much research, where the professor is involved in the process. If a graduate student in his laboratory discovers some new fact, is that wholly the result of the graduate student's brilliance, or did the professor play a role in setting him up? This has sometimes been a source of very serious conflict in universities. Some claim that more than one Nobel Prize was actually the work of a technician, or of a graduate student, rather than the recipient. Fortunately, I never became seriously involved in this conflict. There were students who added my name to their publications, as one of the authors, and there were students who always wanted to publish on their own. I did not contest these decisions, whichever way they went, although I know that some professors refused to have their names on their students' work. I never knew if this was an altruistic motive, to allow the student maximum credit for his work, or if some professors were afraid that their student's findings might not be reliable. I have always felt that there needs to be a strong degree of cooperation in research and, consequently, many of my publications were multi-authored, mostly with scientific colleagues. The drive to "publish or perish" in academia is very strong. Publications affect academic progress, and, perhaps more important, the amount of research money one can attract. Professors, therefore, want their laboratories to be as productive as possible, and this includes students' contributions.

In supervising the work of graduate students, I allowed them considerable freedom in their choice of research. When I was a student at McGill, my professor had favored this policy and I greatly appreciated it at the time. However, in retrospect, I believe that I was probably negligent as an advisor, compared with some of my colleagues. I know that one professor, Dr. Paul Harrison, was constantly concerned about the work of his students and gave them considerable help in their studies. He received a special award for his diligent teaching, and it was well deserved. My feeling that a good researcher should be allowed to discover things entirely on his own was probably not the best approach because of the complexity of many problems and the real need to cooperate. I was, nevertheless, always available to talk, and could be particularly useful toward the end, in supporting a student in his final examinations. Also, it was a great source of satisfaction to me to see my students graduate and track their achievements.

The careers of the twenty-seven graduates I supervised at the University of British Columbia have varied greatly. They range from those who rarely ever used their research skills again (one of my students went into real estate; another became an income tax supervisor), to others who have gone on to start their own laboratories, or to good positions of administration with the government. Having graduate students under your care is not a one-way street; they, of course, contribute much to the total knowledge of any laboratory. At least three of my students, John Parslow, Tom Kessler, and Jim Christian, were excellent at mathematical modeling; several showed great strengths in field surveys; one already had a MD but wanted to expand her experience in ocean science. She eventually became the ship's doctor on a German oceanographic vessel. Another student made use of data collected by an oil tanker that traveled around our coast. He produced an excellent thesis on the waters of Queen Charlotte Sound, which is a potentially important area for oil drilling.

One aspect of training students that I always maintained, as time permitted, was to go to sea with them, or be somehow involved with their field studies. Under pressure to collect samples when it was possible, the students were often at their best, and I was continually encouraged by their work ethic under circumstances that were, at times, difficult. This was when one could chat to students outside of academia and find out more about them. Also, we made various social expeditions together, which included going skiing and hiking in the mountains.

Another teaching opportunity open to many professors is to give a course off-campus, in some totally different environment. In the case of oceanography, this usually coincides with being able to stay somewhere exotic by the seaside and enjoy the ocean shoreline. I gave courses at Friday Harbor (Washington State), in Bamfield on the west coast of Vancouver Island, in Germany at Kiel and Hamburg, at the University of Sun Yat Sen (Taiwan), and several times at the University of Concepcion in Chile. It was a little easier to give such courses in English-speaking countries, but the standard of English among graduate students overseas is generally very high. Only in the People's Republic of China, for a course which my wife and I gave, were we required to have sequential translation, a very tiring process, taking twice as long as normal lecturing. Seasonal courses taught off-campus put us into much closer contact with the students, and this was a rewarding experience. With overseas students, it was also an opportunity for them to investigate studying in Canada.

During my career, I also attended many meetings and seminars to keep up with the latest discoveries. I gained scientific information from these experiences, but I also found myself noticing how the material was communicated. Scientists present their work in a variety of ways and it is interesting to study the speakers as much as their work. In some cases, this adds to an understanding of the caliber of the work, or is an indication of the character of the research methods. Over the years, I came to classify presenters according to certain traits. I am sure that in my own lectures I showed many of these traits, and perhaps someone else has slotted me into one or other of these categories. My personal perspective of speaking is that I probably talk too fast, but also, I have a phobia which has only materialized once or twice in my career. It is to be explaining some scientific point and lose one's train of thought halfway through the explanation—embarrassing, to say the least.

Amongst other scientists, there is that exaggeratedly busy scientist. "Look at me!" he seems to say, as he brushes back stray strands of hair. "I have such a tight schedule! I am in such demand!" He knocks his notes to the floor while trying to establish a particular key point. His visual material is still in transit at the airport. "I'm sorry that my visual material has not arrived," he blusters, "but it just hasn't kept up with me." We are invited to sympathize with the whirlwind progress of such a dynamo.

Some scientists believe that it is their role to be controversial, and the more controversial they are the more they will be noticed. Soon into a delivery by such a scientist, one hears the words, "Contrary to accepted scientific opinion ..." or some similar phrase. This can be followed by open hostility towards certain ideas. It is true that science is controversial, but the best scientists accept the controversy without the hostility.

Then there is the boffin whose ponderous delivery of material in a sonorous voice demands respect for the significance of his work.

By contrast, the pacer rushes to and fro between the podium and the screen. He manipulates a long pointer in a quixotic exhibition of jousting with his illustrations. Once, I saw a pointer lodge in the plumbing of a low ceiling room. I don't think that was in the speaker's notes.

Humor is, of course, sometimes included deliberately. A wry comment or an expressive face can make an otherwise dull lecture more memorable. Occasionally, there is the speaker who has a face and expressions like Danny Kaye, so that even if he is being serious, everyone is entertained by his presentation. The speaker whose jokes are misplaced and lame does the reverse.

Lacking humor altogether are those who feel fellow scientists have not recognized their work sufficiently. Most of their time is taken up with attacks on the work of others. Their disappointments are often compounded by their having just had a grant cut off.

Then there are speakers whose presentations demonstrate imagination

Free to Chart an Independent Course

and initiative, as well as sound research. They make their points succinctly and support them with clearly illustrated data. They leave their audiences with more than an impression of self-importance. I am glad to have listened to many of these.

5. Helping others to navigate

One reason for my coming to the university was to write a textbook introducing students to an holistic approach to biological oceanography. I wanted, first, to express, in general terms, how the physics, chemistry, and biology of the sea were all a single process leading from plankton to the production of fish, and how the fish themselves, as top predators, also controlled the biology of the sea. Secondly, I felt it necessary in such a text to introduce practical aspects of biological oceanography, including the management of fisheries, and problems in pollution.

A young Japanese scientist, Dr. Takahashi, joined me for the project and we came up with a rather small text called *Biological Oceanographic Processes*. It was quite successful, and so we produced a revised edition, which also sold well. I then invited a benthic ecologist, Dr. Hargrave, who worked on the east coast of Canada, to add a chapter on benthic ecology for our third edition, and eventually a fourth. This book was translated into Russian and Japanese, and became a fairly standard text for university students studying biological oceanography in many countries.

Writing a textbook is not without its problems. It is a very tedious process to select material and then to clear copyrights with various authors for the use of their figures and data. This led me to make a bad mistake in the revision of the third edition. When it came to proofreading the final text, I had seen it so many times that I decided to farm out the task to a graduate student who needed some extra money. She read the whole text and only made a few corrections. I thought that the publisher must have done an excellent job, and sent off the corrected proofs. To complicate matters, the publisher, Pergamon Press, in Oxford, England, was having the final proofs printed in India, to cut down costs. When our book appeared, I found to my horror that it was full of many technical mistakes. We eventually corrected these in a revised edition, but I learned my lesson the hard way. After this, I took more care to personally correct the proofs.

We produced another book on oceanographic methods which was partly based on the earlier text on oceanographic methodology, by Dr. Strickland and myself. The new book was made in cooperation with my wife, Dr. Carol Lalli, and a visiting Japanese scientist, Dr. Maita, who made a strong contribution to the chemistry section. We designed the book to be a teaching edition of the earlier book on methods, but it also became well known as a useful text for research scientists. The book was later translated into Chinese.

In the early 1980s, the publishers (Butterworth-Heinemann, Oxford) called me: "We would like you to produce an undergraduate text," they said. "It's for Open University use, over here."

The Open University was originally a UK institution, which has been copied in other parts of the world. My wife and I visited the publishers in England and met some of the people from the Open University.

"We want illustrations, clear accounts of biological processes, not too much mathematics, but above all, end each chapter with a summary and a set of questions for the students," they said.

I suggested that Carol should take primary responsibility for this text. We used some of the information and layout from my earlier text, *Biological Oceanographic Processes*. We worked together, but Carol did all the major composing and arranging of the text, which became known as *Biological Oceanography: An Introduction*. The book was very well organized and appropriate for undergraduate students, of which there was a greater proportion, relative to the graduate levels. Because of its general appeal, it was translated into both Japanese and Chinese, and we sold many times more copies than of my earlier work.

The most recent book that I was invited to participate in was a collection of symposium papers devoted to fisheries oceanography. Dr. Paul Harrison, who had been given the chair that should have been called the Chair of Fisheries Oceanography, invited some notable speakers to talk on this subject, and then asked me to assist in the editing of a book about it. Blackwell Scientific, in Oxford, had agreed to publish the final product. Paul did most of the editing, but was kind enough to include my name. He introduced the idea in the text of having external reviews of each presentation, and these were published along with the original article. This gave the book a balanced approach.

Our contract with the university allowed us to offer our services as consultants with industry or government, for up to one day a week; a generous allowance. This policy was designed to bring the "ivory tower" concept of academia into direct contact with the practical problems of society. I believe that for some faculties, such as Medicine, the consultancy allowance was extended. For the Faculty of Science, it applied to all of us, although the opportunity to consult was very different, depending on the science. In my work on environmental issues, it was quite easy to become involved in concerns such as oil spills or the disposal of mine tailings in the sea. On the other hand, if one's specialty lay in more specific subjects, such as insect taxonomy, there was less chance to consult on the broad aspects of protecting ecosystems. Even further removed from consultations were persons in other faculties, especially Arts, where there was far less money for any kind of consultation than there was with industry.

During my tenure at UBC, I consulted with some thirty companies, including BC Hydro, the logging industry, the mining industry, a panel on offshore drilling, oil exploration in the Arctic, an environmental study of Saanich Inlet, and sewage disposal for Strathcona/Comox, to mention a few. Often, such consultations involved a whole day of dreary meetings with some lawyer, trying to decide the legal difference between a probable and a possible event. I remember a time when the presiding Justice said that anything was possible, but not everything was probable. It was not my business to challenge such trivia, but I would have liked to ask His Eminence to draw me a square circle, or show me two mountains without a valley in between. The legal profession has its own logic. Fortunately, not all consultations were held in a courtroom atmosphere. The further you could get away from lawyers on scientific issues, the more likely you were to reach some reasonable accord.

My longest consultancy was for the Broken Hill Propriety (BHP) copper mine, in Port Hardy, BC. This was an open-pit mine that eventually removed rock down to 1,000ft below sea level. It was located on the shoreline of a fjord, Rupert Inlet. Permission had been given to pump the mine tailings into the floor of the inlet. This was a controversial decision because some people believed that it would harm marine life. The company, which grossed about a million dollars a day from the mine, went to much trouble to demonstrate that this would not be the case. They had a large environmental program, with a laboratory, and boat for collecting samples around the mine site. Samples were sent to other laboratories for independent analysis and government scientists constantly monitored their results.

A précis of the findings showed that the bottom of the inlet was smothered with tailings, to a depth of many meters. Tailings are finely ground rock, together with a small amount of the chemicals used in the extraction process. From independent tests, we knew that young salmon were not harmed by the tailings and, in general, all toxicity tests showed the tailings to be biologically inert. Furthermore, the soft-bottom bed of tailings turned out to be an ideal substrate for colonization by clams and other benthic creatures. I was part of a university team that tried to explain these properties to the public. We had to examine the mine's environmental program, which was described in extensive annual reports. From time to time, we had many suggestions to make to the mine manager, on improvements to the environmental program. In spite of these activities, the mine came under severe attack from the press, which described the mining operation as an ecological disaster, with all the usual jargon being used to take on big industry. Over a period of 22 years, there never was a fish kill in the area due to the mining activity, and the current state of the inlet is a flourishing soft-bottom marine community that is used, commercially, for crab fishing, boating, and sports fishing.

In such studies, I worked in fairly close cooperation with an environmental company called Rescan, whose president is Mr. Clem Pelletier and vice president, Dr. George Polling. From these people I learned some of the problems that environmental companies face in preparing impact statements for industry. These statements are generally designed to ensure the minimum impact of an industry on the surrounding environment. For some projects, statements prepared by Rescan ran to over ten volumes of discussion on environmental issues.

Despite these preparations, there has often been opposition to any new industry. Organizations, such as the Suzuki Foundation and World Wildlife, can get much needed publicity by simply taking on industry, on any issue from fish farming to diamond mining, however well the subject had been covered in the impact statement. These 'green' organizations do not prepare the extensive data sets, discussions, suggestions, and conclusions that are found logically presented in many impact statements. Comments from 'green' organizations can also have a hidden political agenda, in which such subjects as First Nations land claims, encroachment by industry on vacation property, or the support of a traditional fishery may all be enhanced by using the environmental issue to prevent development.

These are often explosive issues in the press, but it has been the role of the environmental companies, working out of the limelight of press coverage, that has provided us with the really important assessments of how to save our environment. Saving the environment should not be a publicity stunt. The problems require logical presentation and careful examination by government agencies, and by university experts experienced in the subject under discussion. The patience of people like Clem Pelletier and George Polling, in endlessly explaining the extent of their environmental studies, has been of great credit to the safe industrial development of our country. Nothing can ever be 100 percent certain in such studies, since each development presents a new set of circumstances. Mistakes have been made in the past, but I believe that in the future we should place our confidence in the environmental companies.

Most of the marine environmental problems, from over-fishing to oil spills, have come to us from the burgeoning and exponentially increasing human population. The same problem exists in the terrestrial environment. A logical role for the 'green' organizations would be to campaign for a reduction in the birth rate. However, they would be unlikely to obtain much popular support for such a mission and this would not ensure their financial continuity.

A very simple calculation on population growth can show how much of the world is losing the battle to improve living standards. Dr. T. S. Lovering of the University of Arizona assumed that the per capita income of a US citizen in 1965 was $2,500, compared with $80 for a citizen of India. The projected Gross National Product (GNP) for both countries could advance at four percent per year. This advance has to be divided by the population growth, however, and for the USA, that might be about 1.5 percent, but it is nearer to 2.5 percent for India. Thus, one hundred years later, in the absence of any inflation, the per capita income in the USA would be over $20,000, compared with less than $500 for India.

While these are all hypothetical figures, they serve to demonstrate how population is such a controlling force in any attempt to solve other problems in the world. The growth of the world population in the next century, toward 12 billion, has all the symptoms of a 'plague' of people. In biology, such populations of other organisms crash when they run ahead of their food supply. Are we any different? Can we prevent this from happening?

While I remain pessimistic about population growth, I am generally optimistic that science will eventually find solutions to global problems, providing the politicians can be persuaded to adopt some possibly unpopular solutions. Religious beliefs and international conflict are other problems that may impair progress toward a harmonious balance between humans and Nature. As a result, I believe that there will be many catastrophes due to food shortages and disease, before humans can live in harmony on this planet.

In the case of the oceans, there have been environmental problems, particularly in coastal communities. The current extent of these problems has generally been assessed, and some remedial action taken. At present, on a global basis, the most severe near-shore environmental impact is eutrophication, or the enrichment of coastal waters with nutrients, either from agriculture or from sewage. This can give rise to extensive ecological change in coastal ecosystems, including the growth of massive amounts of seaweeds of one or two species, or to blooms of undesirable organisms, such as jellyfish. Oil spills are usually the impacts which get the most attention, but oil spills have declined recently, and new threats, such as the use of an antifouling paint containing tributyltin (TTB), are a very severe danger to marine life. We should also consider how much marine environment, in terms of marsh habitat, has been lost for airports, housing, and agriculture, especially in the vicinity of large cities.

Humanity's biggest impact on the marine environment, however, has been the fishing industry, which has decimated so many of the top predators of the sea. There is no form of pollution that has caused as many deaths of marine animals as has this, largely unmanaged, industry. I question whether stocks of fish that have evolved over millions of years, in harmony with the ever changing climate of the ocean, can now survive the massive decimation brought about by fisheries. Part of the evolutionary survival strategy of some species has been to protect themselves in massive schools, which were largely invulnerable to natural predation. However, industrial fisheries are not a natural form of predation, and reducing some stocks to 10 or 20 percent of their original numbers may effectively prevent their recovery. One solution to this problem is to increase our dependence on aquaculture as a source of protein from the sea.

While I have been concerned about the activities of some of the 'green' organizations that may claim to be watchdogs of our coastal environment, there are some who are very sincere and thoughtful. For example, I greatly endorse the work of the Nature Conservancy. They have a program which buys up unique ecosystems, such as some estuarine habitats, for future protection from any industrial expansion. Their work needs to be expanded, to secure larger areas of ocean as protected reserves. Such areas would allow for the re-establishment of fish species that have been drastically depleted by over-fishing. Admittedly, this is a difficult problem, since there is no private ownership of the oceans. Legislation is needed to back up this need for conservation, although it would probably be opposed by the fishing industry.

Sometimes, I was asked to travel as a consultant to another country on some of these issues. Once, I was invited to the University of Pertanian, near Kaula Lumpur, in Malaysia, to make recommendations concerning their marine science program. The university was largely for Moslem students, and the invitation was during the holy month of Ramadan. During this period, the faithful are not allowed to eat from sunrise to sunset. I arrived for my daily interviews with faculty, carrying a small lunch, and I was allowed to eat this in a private room. However, while I felt refreshed to start again after lunch, many of the persons with whom I should be discussing science, appeared lethargic, and lacked any vigor in their opinions during the afternoon sessions. It was like trying to carry on a conversation with someone in a sauna. When I asked one faculty member whether or not he considered his equipment needed to be updated, he replied that if that was needed, Allah would see that it was done.

There were some Chinese Christian Malaysians at the university, but none had been promoted beyond the rank of Associate Professor. My wife and I found this group to be easier to understand, and they invited us to their homes, but we were never invited to the home of a Moslem professor.

Academic consultations in North America have included being part of visiting committees to Scripps Institution of Oceanography, Woods Hole Oceanographic Institution, and the Bedford Institute in Halifax. These are three of the largest oceanographic foundations in North America and the breadth of their ocean science was staggering. Our visiting committees were confined to the biological/chemical side of ocean research in each location. On such visits, it was difficult to recommend what else they should be doing when they were already the leading institutes in the world. If there was a common recommendation for them all, it was to give more encouragement to the younger members of their staff. For example, one of the above institutions had only a single assistant professor on staff, but over fifty full professors.

Another overseas consultancy was in Chile, at the University of

Concepcion. This university has a very vigorous teaching and research program in both fisheries and oceanography. They had invited a Spanish professor to be part of a team of two persons looking at the teaching and research of their academic program. The Spanish professor was Dr. Enrique MacPherson, a strange surname for a Spaniard. However, it seems that a number of Scottish people emigrated many years ago to Spain, in order to help develop the production of port and sherry, of which the British are very fond. They had settled in Spain for good.

As a team of two consultants at the university, we got along very well, and had mostly high praise for the way their science was being done. We did, however, pick up an air of mistrust between students and faculty, when we had an afternoon session with students only. Chile has been through political turmoil in the last decade, and there were many students, and some faculty, who had disappeared or been tortured, during this era. Students were still disturbed and restless on account of these events, and this gave rise to some outspoken comments about how the university was currently run. In general, however, I was very impressed, not only with how the university was operated, but how the whole country managed its affairs.

Compared to many other Latin American countries, Chile is relatively sound, financially; it has prosperous industries, such as mining and fisheries, and it has now emerged as a democracy. In many Latin American countries, any problem with a policeman can be solved by opening your wallet to show your ID, but at the same time, carefully displaying an American twenty-dollar bill. The difference in Chile is that this procedure won't get you out of trouble; it will probably get you the opposite.

Chile has one of the largest fisheries in the world, based on the nutrient-rich water that upwells along the coast. This gives rise to enormous populations of mackerel, sardine, and anchovy, together with a rich bottom fauna. Marine predators, including large bird and sea lion colonies, are on islands near the coast. We went out to sea from Antofagasta, to visit one of these bird and sea lion colonies. We left early in the morning, when the waters were covered in a deep fog. The boatman seemed to know where we were heading, although it was impossible to see more than a few feet ahead. You just had to follow your nose. The dried excrement of these colonies forms guano, which is still mined as a fertilizer on some islands. As we sped through the water, the smell gradually became overpowering.

We broke through the fog about ten miles off the coast, and in front of us was an archipelago of white-domed islands, with large numbers of Humbolt penguins, red-footed boobies, turkey vultures, and sea lions basking in the hot tropical sun. Our boatman quickly took us to the windward side of the islands, or we would have been asphyxiated by the stench. Although the rocks were very rugged and steep, the sea lions make a habit of climbing up as high as possible, sometimes right to the top of a jagged peak. Mothers will even haul their young up over the steeper terraces. Some sea lions fall on the way down from these lofty heights and that is when the turkey vultures get their meal.

There are other wonderful natural sites in Chile that are on few tourist maps. One of these is the Atacama desert, east of Antofagasta. It is the driest desert in the world, and whole generations of people can grow up there without experiencing rain. In Antofagasta we boarded a very modern "Tur" bus for the regular daily run to San Pedro. After winding through a few city streets, the bus started off in a straight line across the desert, on a well-paved road. The elevation increased gradually from the coast, to about 8,000ft. The countryside was virtually stripped of all vegetation; just unending red, sandy dirt for miles. Occasionally, there was a building, which might have been part of an abandoned nitrate mine, but otherwise, it was sun and sand for a couple of hundred miles. It all looked like a close-up image of the surface of Mars.

Once in a while, our peaceful ride was shattered by an alarm going off in the bus. From the electronic speed indicator above the driver's seat we could see that this always occurred at speeds of over 90 km/hr and resulted in the driver slowing down. We reached the verdant oasis of San Pedro de Atacama in about four hours and booked into the abode-style Hotel Tulor. The town of about 1,000 inhabitants had dusty streets but a general ambiance of total relaxation. The brown mud walls of the houses were punctuated with gaily colored doors painted red, blue, and yellow. Little restaurants abounded up and down the street, and the food was both excellent and inexpensive. The local meeting place was around the Catholic church, which was built in 1577.

San Pedro is very close to the borders of Peru, Bolivia, and Argentina. From here, we took day trips to elevations up to 14,000ft, where the snows of the Andes and bubbling hot springs provide enough moisture for many plants and animals. The best geysers are found at El Tatio, and here I was able to have a swim, although the outside temperature was freezing and the altitude did not allow for vigorous activity.

Two distinctive animals of the Alto Atacama are the guanaco and the flamingo. The guanaco is the only member of the family Camelidae that is not domesticated, although the wild animals are herded annually by the indigenous Aymara people, for their wool. As we drove through the high desert, what had seemed to be boulders in the distance, suddenly started to gallop across the plain with great elegance. They stopped just as suddenly as they started, and peered back at us with a 'catch me if you can' attitude. The flamingos are found on the Salar de Atacama, where they feed on brine shrimps; there are three species and all of them are generally found more abundantly in Bolivia. They stood on their high stilt-like legs with their beaks submerged, swaying their heads from side to side as they filtered out the tiny crustaceans. This was all back-packing country, undiscovered by large tourist groups. It was a wonderful adventure to have included in my visit to Chile.

In 1998, my wife and I visited the Galapagos Islands. The plane landed on San Cristobal Island and after paying the park fee of $200 US (in crisp clean notes, only please!), we were ferried to Peurto Ayora on the island of Santa Cruz. From there we took a brand new tour boat to visit other islands. The tour started out well and at first I was very excited with the fauna and flora. However, after about three days of trudging through dry lava beds and desiccated vegetation, I became a little bored. It was the year after an El Nino and the effect of this had been to starve many sea lions. Their corpses littered high and low ground, having struggled there from the sea to die. The surviving sea lions were a great attraction and they hitched rides on all the smaller boats. The females would cavort with swimmers in the water, which was fun, as long as there were no aggressive males around. The snorkel diving was not as good as in some other rock and coral locations in the world but all the coastal fish, including the abundant sharks, seemed to be larger here than elsewhere.

Our trip on the tour boat was cut short when, after visiting several islands, we were the only tourists left on board. Within hours, our vessel was invaded by about twenty ex-army Israelis on a grand tour after their national service. They had been sold the accommodations by a local Israeli travel agent at a reduced price of $200 per person, while our fare was around $1,000. They were a boisterous lot who grabbed all the deck chairs, ate all the buffet food, and spoke to each other loudly in Hebrew. We demanded to be put ashore and to have our fare returned. Fortunately, the tour line agreed to both requests. We stayed several nights, instead, at a small hotel in Peurto Ayora. This was a delightful experience and gave us a chance to visit the Charles Darwin Research Center and see the only giant tortoises on public display. It would have been better to see these animals in their natural habitat on Isabela Island, but this interesting island with its active volcanoes is not generally accessible to tourists.

In 1991, I spent a couple of weeks in St. Petersburg, attending a world symposium on ocean pollution. Wonderful sights in this city include its golden-domed buildings, which shine like stars in an otherwise dreary landscape, and, of course, the Hermitage Museum. Built in about 1750, the Hermitage, formerly the winter residence of Russian czars, has a gilded interior and houses over 3,000,000 works of art, from Neolithic artifacts to modern paintings. The downside of St. Petersburg was the suffering of the people who had little to eat, the drab appearance of every building, roads with potholes big enough to fall into, cars breaking down all over town, and the endless line-ups for the necessities of life—a queue of forty cars at a gas station being a common sight. Hotel food seemed to consist mostly of rye bread and cabbage—raw cabbage, cabbage soup, boiled cabbage, cabbage patties, and cabbage-a-la-king. Also, I did not feel safe in St. Petersburg. Many Russians were openly hostile to westerners.

I took some cans of ham with me to Russia, to give as presents to other scientists. However, I found that the scientists were much better off than the average citizen. I met an old lady who cleaned toilets in our hotel. We communicated in German, since she had been a German translator in the Red Army. When she retired, she said it was like a death sentence—her pension did not allow her to do anything. So now, as an aging grandmother, she had to clean toilets. I gave her a can of ham and some bars of chocolate for her grandchildren, to supplement their cabbage diet.

Another country where cabbages play a central role in the diet is South Korea. As a main ingredient of Korean cooking, it is known as Kimchi, and is highly spiced with hot pepper. In fact, most Korean food seemed to be just blazing hot, without much refinement of spices. I went to Seoul to give a plenary lecture at the Pacific Science Congress; we were generously hosted by the scientific community, but outside the meeting rooms it was a different story. As a divided country, there is great underlying tension in the population. South Koreans appear as a tough bunch of complainers, and while I was in

Seoul, staying at the Hotel Lotte, the staff went on strike, along with workers from several local industries. Striking workers brought a lot of politics into their complaining, and it almost seemed as if the loudest complaints were from pro-North Koreans implants. Despite having one or two good friends in South Korea, I did not, at any time, feel comfortable in their society.

Apart from these accounts of service and travels overseas, there were also a number of scientific service functions to be undertaken on the home front, such as editorial work for journals and scientific evaluation committees There were also opportunities for sabbatical leave and taking part in promotions and awards of one's colleagues. All of these were often rewarding experiences.

Being invited to be on the editorial board of an oceanographic journal was an honor I usually accepted. The workload was light, requiring me only to review several manuscripts by other scientists, each year. When I became Editor in Chief of the new journal I had helped to found, I came to realize more fully what was involved in putting out a new number; adjudicating on reviewers' comments, tactfully turning down some manuscripts, and finally, dotting all the i-s and crossing all the t-s before sending an edition to the publisher. Sometimes editing had its amusing side.

One author wrote in, "I would like to have Smith and Jones to review my work. They are the only two people who really understand what I am doing." I duly sent a copy of the submission to Smith and also to Jones. Both referees turned the work down for various reasons and suggested that it should not be published.

"I'm sorry that I am not able to publish your paper," I told the author, "but neither of the independent (anonymous) reviewers recommended its acceptance."

The author wrote back, "I asked you to send my work to Smith and Jones. I knew it would be turned down by any one else."

The journal that I was editing was *Fisheries Oceanography* and it was part of my effort, discussed earlier, to bring a new science on line for fisheries management. Initially, scientists were very reluctant to use a new journal, and my first number was mostly by invitation to prominent scientists to write an article, in order to establish some respectability for our publication. The number was successful, but then there was a scramble to get sufficient quality papers for the second and third numbers. Once the journal was accepted as one that would appear on a scientist's citation record, then papers started to flow in faster than I could publish them. The citation process is determined by a published index (*The Science Citation Index*), which lists the journals that are scanned by different authors for publications to cite. The number of times an author's paper is mentioned by other scientists is taken as an indication that the scientist's work is being recognized. Thus, in establishing a new journal, it was important to have it as one of those which was scanned for authors' contributions.

Apart from my role as a consultant for private enterprise, I was sometimes asked to serve on National Research Council committees for the distribution of funds to researchers. This was part of a democratic process in which we, as applicants for the same funds, were asked to serve for some time on the committees that distributed the funds. Our own research grants were either frozen for that period or adjudicated on when we were absent. The work was exacting. We had to read perhaps one hundred research proposals, but it gave us a good insight into what else was going on in the research world.

The most interesting awards committee that I chaired was for the award of 1967 Centennial Scholarships. These scholarships for graduate students were worth much more than the normal NRC scholarships, and we interviewed only the brightest and the best of Canadian students. I was particularly interested in talking to these students to find out where their interests lay. They were all very articulate and quite outspoken.

"I hate Christians," one student said. "They have caused so much trouble in the world."

Another said, in response to a question about his future career, "Well I can tell you, firstly, some organizations that I will never work for and they are the charitable ones. Charity takes away the freedom of spirit to do things for yourself. The welfare system has produced a class of zombies in our society."

Some common characteristics emerged almost as a pattern among them. As a group they were scientists, but nearly all of them had some musical talent, they read books which were either science fiction or fantasy, they were often interested in some invented language for communication, and they often had an interest in participating in (not watching) sport.

Sabbatical leave at the university was generously allowed. The rules changed now and again, but while I was at UBC, we were allowed six months of leave at seventy-five percent salary, every four years. Some persons took full advantage of this and went off to another institution as often as was allowed; others never took sabbatical leave. This was not a good idea because the absence from local matters and the stimulating company of different scientists were good ways to refresh one's thinking. I went on sabbatical leave twice, although I was also away at other times, such as during our research program in China. In 1978, I went to Australia, where I had been corresponding with a scientist at a government laboratory in Sydney.

During that leave of absence, I developed a hypothesis on the origin of food chains, based on evolutionary ideas that I have described earlier. In 1986, during another sabbatical in Plymouth, England, my wife and I wrote a review of the differences between the ecology of the North Pacific and North Atlantic Oceans. Although these two oceans are situated at the same degrees of latitude, they have very different ecologies, and so what scientists were reporting for the Atlantic Ocean might not be applicable for the Pacific Ocean. This exercise in comparative ecology seemed, to me, an important way to show how two areas, though geographically similar, could be ecologically different.

One curious aspect of our sabbatical at Plymouth was the obvious stratification of scientists and technicians during coffee (or tea, in England) breaks. I was accustomed, in Canada, to mixing with all employees during such brief moments, in order to learn what was new in anyone's life, scientific or domestic. During such daily meetings, we often spent much of our time around a blackboard, where we could outline our latest ideas. This was not the case at the Plymouth laboratory, where there was a rigid hierarchy of where you sat to sip your beverage. The most senior scientists were seated at one end of the room, with the most junior technicians and secretaries at the other. At the National Institute of Oceanography, this seating plan was taken a step further; the director actually had a little curtained table for added privacy. My wife and I were not sure where to sit, but were generously encouraged to sit somewhere near the more senior scientists. I emphasize that this was a peculiar arrangement because I know that many persons, including myself, regard the coffee room conversations to be very important exchanges for research ideas. Sometimes these ideas can come from very junior persons. In spite of this, we got along well with a number of colleagues, including a mathematician, Arnold Taylor, who helped me develop ideas on modeling, a zoologist, Roger Harris, who was later organizer of a very large international zooplankton program, and Robin Williams, who was generous in helping us with the interpretation of data.

Living in England for six months in 1986 brought back some contrasts and memories of the country that I left in 1949. My grandfather's church

Free to Chart an Independent Course

was still strongly Anglo-Catholic, but there was now only one priest for four parishes, so services could no longer be held at St. Bridget's every Sunday. The road traffic in Britain had become terribly congested. I bought a car, but halfway up to Oban, on a visit to another laboratory, I tried to sell it after many hazardous encounters around Manchester and Liverpool. I have driven the busy streets of Paris and Los Angeles, but the unrelenting pace of the bumper-to-bumper traffic of the English Midlands drove me to despair. I was unable to sell the car at that time, but curiously, on returning to Plymouth, it was practically demolished, one night, by a drunken driver, while securely parked in front of our apartment.

English food has greatly improved in the past forty years and it was a pleasure to find continental cuisine and Asian curry houses all over the place. The English pub was still a wonderful institution, but in Plymouth, every weekend, a free-for-all developed at closing time on the street outside one pub near to us. The noise of police cars, yelling hooligans, and, eventually, ambulance sirens, always occurred for at least two hours every Saturday night. There was never any logical explanation for this behavior. It was just a rather dangerous local game that some patrons engaged in every week. This seems to happen in North America too, where there is often no clear motive, just a lot of rather young and energetic people wanting to have a crack at authority.

One of the more enjoyable services I undertook was assisting in obtaining awards for others. Oceanography is not a field in which there are many awards, but I was active in initiating the election of scientists to the Royal Society of Canada, promoting the name of someone for a special medal, and supporting other persons' ideas on who should receive greater recognition. I was about seventy percent successful between nomination of a candidate and his or her award. Yet still I feel that there are many persons in science whose outstanding careers have remained largely unrecognized, while politicians and sports persons are displayed every day in our newspapers as icons of our society. Babe Ruth and Wayne Gretzky are household names, but who invented the World Wide Web? (Before you scramble for a Google search, it was a graduate of Oxford University, Timothy John Berners-Lee. He was awarded a Japan Prize in 2002).

I greatly admire, too, the independent work of other oceanographers. I could write endlessly about other people's science, but as an example of a person whose work I hold in high esteem, I will choose Robin Pingree of the

United Kingdom. In 1978, Robin published a paper about tidal fronts and how photosynthesis maximized at these particular locations. It was an excellent piece of work, which was done partly in cooperation with a physical oceanographer, Dr. Simpson, also from the UK. A student of mine the time, Ian Perry, later made very successful use of Pingree's work in his analysis of the waters of Queen Charlotte Sound, where I had my first encounter with oceanography, many years earlier. This was an example of how other scientists sometimes helped my students to navigate in their own research efforts.

It was after my sabbatical leaves and when I had decided to retire early, that I received a couple of prestigious oceanographic awards. In 1990, I received the J. P. Tully Medal from the Canadian Society of Meteorology and Oceanography, for my work on fisheries oceanography. The ceremony was particularly special, since it had been the late Dr. John Tully who had given me my first job. In 1999, I received the Murray Newman Award from the Vancouver Aquarium. This was associated with my work on the coastal waters of British Columbia.

These were all wonderful surprises for someone who had enjoyed doing what was almost a hobby, but which seemed to generate at the same time, some outside recognition. Shakespeare used a curious word to describe one who obtained many honors; *honorificabilitudinitatibus* (*Love's Labor's Lost*, Act V, Sc.1). Not the kind of word that you want to bring up in general conversation, but it does reflect a little on the fact that scientific honors may be given disproportionately to a few people, while other excellent scientists seem to be passed over, unnoticed.

6. The Northwest Passage

The Canadian Arctic Archipelago extends from 70° to 80°N and covers an area larger than any of the Canadian provinces. It is a vast land, stretching into an almost limitless hinterland of pristine landscape, broken only by rugged mountain peaks. When the thin rays of sunshine fleck its surface, ice crystals make diamond-like sparkles in fleeting patches and ever changing patterns of light and shade. The deep blue waters are constantly smacking against the ice edge, and the penetrating sound of cracking and groaning ice can be heard all around.

Within this area, the sea water circulates approximately from west to east. These are the waters of the fabled Northwest Passage, where Sgt. Henry Larsen of the RCMP was the first man to sail a vessel, the *St. Roch*, from the Pacific to the Atlantic, in 1942. His return voyage in 1944, from the Atlantic to the Pacific, was the first time anyone had crossed in both directions. The information gained on his journey opened up significant questions about Arctic oceanography. These are particularly important because of the anticipated melting of a lot of polar ice within the next century. Some foresee that commercial vessels from Europe and Asia may one day use this seaway as the shortest route between the two continents.

The passage is about 4,000km and threads its way from the Beaufort Sea, through McClure Strait, Lancaster Sound, and into Baffin Bay. The oceanography of the area is still poorly known. This was brought home in the 1970s and 80s with the discovery of oil and gas reserves in the Beaufort Sea. Before further exploitation of the area, we need to know more about the environmental impact of using the Arctic Ocean for commercial activities.

I made my first visit to the Arctic Ocean in 1980, as a consultant for Dome Petroleum, who were drilling for oil on various kinds of platforms in the Beaufort. The one that I visited in the middle of winter was a large Japanese tanker grounded on a gravel shoal that had been dredged up for that purpose. The drilling platform had then been built through the stern of the ship's hull, and about fifty or more employees were working twenty-four hours a day, drilling for gas or oil through the floor of the ocean.

The Arctic and the Sahara have something in common. They are both deserts. However, neither is completely devoid of vegetation and animals. In places, both the sub-tropical and arctic deserts are bordered by areas of highly productive ocean waters, which give rise to a myriad of life forms. The Arctic in mid-winter is a time of continual darkness, but the white reflection of the ice seems to cast a dim glow over the landscape. The displays of northern lights add an eerie iridescence of light and shadow over the whole landscape. The moaning wind, an occasional loud sound of cracking ice, and the darkness give one the impression of having landed on some other planet.

After arriving in Inuvik, I transferred, in a small plane, to Tuktoyaktuk, which was the base for Dome Petroleum. A helicopter took us offshore to the ship. Half the tanker had a hotel-like atmosphere, with all sorts of amenities, everything, in fact, except a bar. Alcohol was strictly prohibited on the base. The other half of the tanker was a machine shop, including the drilling platform. Everything was indoors, save for a lonely Inuit who patrolled the perimeter of the vessel with a number of dogs. Employees of the oil company were allowed to go off the vessel onto the ice, but this could be dangerous if there were polar bears around; these are most active during the Arctic winter, when they go out on the ice to hunt seals. The lonely Inuit was the polar bear watch. The hairs of his face were frosted under the hood of his *atiigi* and he stomped on the ground to keep his feet warm in his sealskin *mukluks*. A rifle was cradled in his arms and he was ready to fire a warning shot, if the dogs scented a bear near by; though attracted by human habitation, polar bears were easily warned off by the patrol.

In consulting with Dome, I was to comment on the biological aspects of oil spills in the Arctic and to suggest mitigative measures for their control. I had seen two large oil spills earlier in my career: the super tankers, *Torrey Canyon* and *Amoco Cadiz*. Both had grounded in the English Channel on the English and French coast, respectively. The damage from these oil spills, of over 200,000 tons, caused devastation over a coastal area of approximately 100km. After about ten years, there was little superficial evidence of the spills; both were in areas of high wave activity. The Arctic, with its winter cover of ice, was a very different case. Oil could remain in the area for decades, and cause a lot of damage to the bird life, seals, and also to Beluga whales that tend to congregate close into shore for breeding.

We held a public meeting in Inuvik for the inhabitants to express some of their concerns, and I was hard pressed to satisfy the questions that followed. The audience was divided between those who wanted the jobs that the industry created, and those who saw their way of life being eroded by such a vast army of potential intruders. Most of the questions came from the latter group, and they expressed genuine concerns about the damage that an oil spill could cause in the Beaufort Sea. At one point in the discussions, I used my experience with the Aswan dam as an example of how environmental impacts can sometimes become grossly exaggerated. This caused the Mayor of Inuvik to challenge my whole background experience, on the basis that I was too young to know what I was talking about. I was greatly flattered to be fifty and too young! I gave her a copy of my résumé, which listed my experiences, and we actually got along quite well afterwards. Her concerns were well justified regarding the hazards of an oil spill in the Arctic. The best defense was to rely on the company claim that an oil spill would not happen under the stringent drilling controls imposed by government, but such assurances are never certainties.

My second and third trips to the Beaufort Sea were made on the Department of Fisheries and Oceans research vessel, *John P. Tully*. This is a very well-equipped vessel of approximately 1,500 tons, with a reinforced bow for breaking thin ice. Once on board, I discovered a very pleasant dining room and lounge, and an exercise room with a bicycle, weight lifting equipment, and a ping-pong table. The food on board was excellent, and the scientists ate with the officers. There was a hot tub on the ship's deck where we could enjoy a warm soak at any time of the day. Sometimes, we sat in it at midnight when the sun still shone, soaking in hot water, on our own little floating world, surrounded by a frozen landscape.

Despite this modern setting, the captain was both disagreeable and old-fashioned. We had a satellite navigation system on board that was malfunctioning, so the makers of the equipment came out on a helicopter from Tuktoyaktuk and fixed the settings. The first officer was very pleased with the result, but when he reported to the captain on the bridge, the captain's comment was, "Turn the damned thing off! Do you think I don't know where we are?"

Apparently, he was unwilling (or did not know how) to use the equipment, and considered it a challenge to his knowledge of other navigational aids. He reminded me of some of the fisheries scientists that I have met who are unwilling to change the science that they were taught in university thirty years ago.

I came on board the *Tully* by helicopter, out of Tuktoyaktuk. Our mission was to learn something about the processes of biological production in the Beaufort Sea. The area is a very productive water body in the summer months, and Bowhead whales travel all the way in from the Bering Sea to feed in the Canadian Beaufort. During the winter, this sea is covered with ice and the only ice-free water occurs in isolated pockets, known as polynas, along the junction of the land-ice and sea-ice, some distance off shore. These are important areas for seals and walrus in the Arctic. In the summer, the ice breaks up and the waters produce an abundance of plankton. For the waters to be so productive, it is not hard to realize that there must be adequate sunlight and nutrients. The sunlight is no problem because there are twenty-four hours of that within the Arctic Circle. The sustained input of nutrients, however, is not immediately easy to understand. Deep water in the ocean contains lots of nitrates and phosphates, but they must be brought up into the sunlight zone (the euphotic zone). There were two possible mechanisms for this in the Beaufort Sea. Either the nutrients were coming from the Mackenzie River as it entered the sea, or there was some upwelling occurring along the continental shelf of the sea.

We designed a research program to find the source of the nutrients. We measured nutrients in the water but, more importantly, we measured the ratio of heavy and light isotopes in the plankton of the sea. By measuring the quantity of heavy and light nitrogen and carbon isotopes, which occur naturally in the plankton, we were able to tell if the animals were deriving most of their food from the river or from the sea. It turned out that only one group of small crustaceans was heavily dependent on the river for sustenance, but that almost everything else we measured depended entirely on the sea for nutrients. The next question was how the nutrients were being brought to the surface along the continental shelf of the Beaufort Sea.

The answer came from a lecture I heard by a physical oceanographer, Professor Susan Allen, who explained how seawater was forced to the surface from great depths when there was a current flowing over a canyon on the coast. In the Beaufort Sea there is a very large canyon, the Mackenzie Canyon, that is positioned on the continental shelf near the river, and then another, smaller canyon, some distance to the east. It was our belief that it was the position of these two canyons, coupled with the west to east flow of the Arctic current, that supplied the nutrients needed for the massive blooms of plankton.

Before going to the Arctic, we knew about the species of animals that lived there, and about their life cycles. What we did not understand was how the whole ecosystem worked. By carrying out our experiments, we had solved a small piece of the puzzle, in regard to the mechanism controlling production in the Beaufort Sea. This kind of information is of vital importance, if we are to understand the Arctic marine ecosystem before it is used by commercial vessels, which may be sometime in the next 100 years, assuming the current rate of global warming continues.

My last trip to the Arctic was not for oceanographic research, but to advise on an environmental recovery plan for the Ekati diamond mine. In one of the greatest mining adventures of all time, Charles Fipke, with some assistance from others, discovered diamonds in the Canadian tundra in the late 1970s, and a mine, operated by BHP Billiton, was eventually located about 300km northeast of Yellowknife, near Lac de Gras, in the Northwest Territories. The mine is most successful, and currently produces about 50,000 carats of diamonds per day, in a mixed array of industrial and gem quality diamonds.

The diamonds are contained in grey kimberlite pipes, which are ancient lava plugs in the earth's surface. This rather soft mineral is generally worn away, relative to the surrounding rock, so that the sunken pipes are usually found under lakes, which have to be drained. Open pits are then dug and the kimberlite extracted. As a result, several large holes, about 1km wide and 500m deep, have been created around the mine. The environmental recovery program for these pits is to turn them back into biologically productive lakes that will once again meld with the rest of the tundra environment. However, the dimensions of these new lakes will be very different from most of the existing lakes in the area, and some thought has had to go into designing a recovery program. From my earlier work on Great Central Lake, and my involvement in the recovery of the copper mine at Port Hardy, BC, I accepted an offer to advise on the BHP Billiton recovery program.

This environmental work is exciting because the whole ecosystem is so different, being located above the tree line and in the permafrost. Great herds of caribou crisscross the countryside, pursued by wolves, wolverine, and barren-land grizzly bears. The bears are almost entirely carnivorous in these latitudes, and their technique for catching the much swifter caribou is ingenious. The tundra consists of many small lakes, boulder fields, and vegetated areas of scrub bushes and grasses, on which the caribou graze. The bear's attack is to ambush and frighten the caribou, so that in their sudden panic, they end up in a boulder field or a lake. In both environments the bear can then move faster than the caribou. This is particularly true on the boulder fields, which caribou otherwise avoid because they can break their legs. They become an easy prey for bears under either circumstance.

The lakes have a number of species of large fish, such as grayling and trout. These are important to sports fishermen, and form the basis for commercial fisheries on the larger lakes. For six months of the year the whole tundra becomes alive; it seems a great mistake to call this area "the barren lands." The lake recovery program will add a very small, but somewhat different ecosystem to this area. In general, mine site recoveries range from doing nothing and leaving a mess, to the famous Butchart Gardens in Victoria, BC. The latter was an open-cast mine that was turned into a world class botanical exhibit. The recovery of the Ekati mine site will be effective, but unlikely to compete for any botanical prize.

7. Gumboot oceanography

Gumboot oceanography does not take you out into the oceans, but it is one of the best field exercises for student training. Oceanographic ships have a very limited availability and, consequently, getting students out to sea is difficult. If they want to repeat a study, they may find that there is no ship available at another time. Putting on their gumboots and marching along the shoreline can be done at almost any time, and with much less cost than for an oceanographic ship. If one only has a limited time in which to complete the requirements for a graduate degree in marine science, a good pair of gumboots might help.

As a result, although most of my oceanographic studies involved the ecology of the water column (or the pelagic environment), with limited time and financial resources it was sometimes easier for some of my graduate students to study near-shore communities. Studies of organisms living on the ocean floor are referred to as benthic ecology. These two ecologies (benthic and pelagic) merge together in the littoral zone, where the sea meets the shore. A few shorelines are well known for their sandy beaches and excellent bathing, but the vast majority are either rocky outcrops, massive forests of mangroves in tropical waters, or marsh grasses in temperate waters. These areas are of particular interest to marine biologists because they are almost always the most severely impacted by human development. For example, urban development of coastal cities such as Tokyo, New York, and Vancouver has resulted in the loss of more than 90 percent of the natural ecology of the coastline, and its replacement by airports, housing, and agriculture. Such flat areas are an easy target for developers who drain, dike, and fill-in marshland for resale at premium prices. Some of my research was carried out in these near-shore environments.

My earliest experience of sampling the shoreline was to collect some random samples of plants and animals. We assured the randomness of our sampling by throwing an aluminum "picture frame," one meter square, onto the sediment surface, and then sampling everything that was captured within the frame. I hurled my first frame far out across the beach. It hit a rock and shattered into several pieces. At least the plants and animals in that square meter were spared an ignoble end in my formalin jars.

Another time, we were laying out transect lines from the high to low water marks, when a big black bear emerged from the bush and started to sniff his way towards us. Fortunately, our lunches were still in the car, so we retreated and waited. The bear wandered out to the low water mark, where he was seen to be licking some rocks. Since that was too far to go for a salt lick, we were curious to find out what the attraction was. After he left, we found that the rocks were covered with large numbers of amphipods (some species are called sand fleas). These little animals must have been a kind of caviar to his palate.

On the Fraser River mudflats, we conducted a survey for mercury in all kinds of animals, including mussels, clams, and crabs. In an area closest to the municipal sewage outfall, the level of mercury in all of these animals was greater than the one part per million (1ppm) allowed for human consumption, but at the same time, the animals were apparently healthy. The explanation for this incongruent result was that they all contained high levels of a protein called metallothionein, which can be produced naturally by most animals as a protection against heavy metal poisoning. While sudden doses of mercury could be toxic, over time, these animals had developed a protection to elevated mercury levels coming from the industrial component of the municipal outfall. Interestingly, the whole rocky ground around the outfall was the most active biological community in the area. In addition to the abundance of marine organisms, there were many rats, feral cats, snowy owls, and hawks. Of these organisms, we measured mercury only in the rats, whose diet consisted, in part, of mussels. Again, they had high mercury levels, which they probably passed on up the food chain to the larger raptors.

Apart from the presence of mercury in the sewage, we wondered if mercury could also enter the marine food chain from natural deposits. On the west coast of Vancouver Island there is a cinnabar ore deposit which extends into the sea. The reddish rocks are covered with turban snails grazing on the epiphytic algae. We collected some for analysis, but found that the mercury levels in our samples were normal. Mercury from the cinnabar ore had not become mobilized into the marine food chain. The research of other scientists has shown that mercury in seawater enters the food chain because of the large quantities of organic matter in sewage. This is metabolized by bacteria, which also take up the mercury and pass it on to filter-feeding organisms, such as mussels. The mercury ore had no such supply of organic matter. These results are counter-intuitive, in the sense that one might suppose that any source of mercury in the marine environment should be considered toxic.

The Sea's Enthrall: Memoirs of an Oceanographer

Slogging around in thick mud is very tiring. Mud sucks. Each step is taken with great effort and it is often better to wear something equivalent to snow shoes, in order to get around. Mudflats can extend for many miles out to sea. This presents a certain hazard for workers trying to collect samples during low tide. Once the tide turns, it can rush back over the flat ground with a vengeance, trapping any overly zealous scientist who has become too wrapped up in his sampling program. The best solution to this problem has been to use a hovercraft, which can quickly transport the scientists over the muddy substrate, or the water, with equal ease.

Most intertidal areas below the high water mark are considered public property and accessible for scientific study. This is not true, however, in some states in the USA. I was working on the shoreline with students out of the Friday Harbor Marine Laboratory, on San Juan Island, in the state of Washington. A local waterfront owner, at some distance from us, let loose with a firearm to gain our attention—a warning that we should get off his property. In Washington, that meant right down to the low water mark. At some time in the past, this state, regrettably, sold off these rights to raise some extra cash.

In British Columbia, the logging industry is one of the principal sources of income for the province. Consequently, it has a powerful political lobby, and this has been used to ensure production of lumber at the lowest possible cost. One cost-effective measure has been the ocean transportation of logs, tied together in enormous rafts. These are assembled in the near-shore waters of the logging camps, and are towed to the near-shore waters of the timber mills. Both sites can be severely impacted by this practice. The weight of the logs, which are often grounded at low tide, and the continual deposition of bark debris, destroys most of the natural ecology of such areas. I worked on a committee for nearly a year, in which we tried to come up with alternative methods of log storage and sorting. Many of the areas impacted were formerly feeding grounds for juvenile fish, particularly, young salmon.

The outcome of our work was largely a compromise between industry and the environment. There is seldom an absolute solution for such problems. Our industrial needs are a part of human evolution in the natural world, of which we are a part. Nature can be just as severe as humans in destroying some coastlines, e.g., by erosion, or by rearranging whole ecosystems with hurricane force winds. It is a misconception of newspaper articles to refer to ecosystems as "fragile." Nature is inherently resilient, or it could never have survived global evolutionary changes in climate and geography.

At present, the most severe form of near-shore coastal pollution throughout the world is eutrophication, or the nutrient enrichment of near-shore waters by sewage and agricultural runoff. One indication of this is a mass of green algae on the foreshore. This is indicative of changes in the whole littoral ecosystem, and the problem is far more persistent than any oil spill or other forms of coastal damage. How to fix this problem is a challenge for the twenty-first century

There is a particular area of the coast of British Columbia called Baynes Sound, which is famous for growing the best oysters and clams. Some years ago, I suggested to a PhD student, Mike St. John, that he might write a thesis on why these creatures grew so well at that particular location. He accepted the challenge, and his thesis showed that in a nearby body of water, known at Lambert Channel, the movement of the tide through this narrow passage brought nutrients into the euphotic zone throughout the summer months. This channel was a "tidal front" which produced large amounts of phytoplankton, with a periodicity of about fourteen days, matching the neap and spring tidal cycle. This information became rather important about twenty years later, when the shellfish growers wanted to expand their operation. Shellfish culture is no different from other kinds of farming, in that you can only put so many ponies in a paddock before you run out of grass and the landscape is covered in feces. There had to be some measure of the carrying capacity of Baynes Sound for shellfish culture, and the work done much earlier, by my student, was important to this study. Mike St. John is now a professor of oceanography at Hamburg University, in Germany.

Two of my students, Denis D'Amours and Don Webb, studied the feeding of young salmon in the Fraser River estuary. Although there appeared to be many kinds of small, crustacean prey items in these areas, the young salmon were very discretionary and fed almost entirely on a single species of benthic copepod. The timing of the abundance of this particular prey item was critical to the arrival of the young salmon. These times had to coincide, for maximum salmon growth and survival. However, these two events were not always synchronized. It was not part of my students' research, but one has to wonder why the young salmon chose one particular prey item, when there seemed to be so many other kinds of small crustaceans available. This is probably a question common to economists, who are constantly analyzing what makes one item so much more successful in the market place than its competitors. Convenient size, nutritional content, palatability, abundance, and ease of capture must all register in the genetic response of the young salmon.

The two graduate students in these studies had to carry out some of their sampling during the spring, in the middle of the night. This was because the very low tides which uncover the mudflats only occur at night, during the early spring. The tidal waters can flood the area very rapidly, so naturally, this night-time sampling schedule was an added concern for safety. Many ecologists have had similar dangerous experiences associated with their work. This may be why some of the best ecology is often done by young scientists and why older scientists prefer the comfort of their offices.

Some of our near-shore work involved comparing methods with scientists in other countries. In Germany, I was invited by Professor Krey, one of the pioneers in ocean chemistry, to take part in some intercalibration experiments at sea. We went on board the research vessel *Hermann Wattenberg*, in order to collect samples from coastal waters. The captain of the vessel was referred to as "Herr Ober." His role seemed to be to get the vessel into international waters as soon as possible, so that he could distribute duty free liquor. This meant traveling only a short distance offshore because of the close proximity of national boundaries in the narrow region of the Baltic. After we had carried out our experiments, several bottles of "Duty Free" were opened, and the day was concluded in a most agreeable manner.

Free to Chart an Independent Course

PART THREE

In Dry Dock

IX
Life in Home Port

My involvement in a scientific career entailed much time away from home traveling, attending conferences, or just spending late nights in the laboratory. As a result, my domestic life suffered from a lack of attention. I do not know how one strikes a balance between a dedicated career and a desire to be a functional part of a family. With different genes, others have found better, and sometimes possibly worse, solutions to this dichotomy. Life has become increasingly complicated, compared with that of our hunter/gatherer ancestors. While these complications have been generally recognized, how to deal with them is not a matter of "reading the manual."

At the time that I attended an English boarding school, teachers were not known for instilling into their pupils any concepts of family life. With the death of my father, when I was very young, and after spending seven years at a boarding school, where I was out of contact with family affairs for nine months of the year, it is not surprising that I really had very little training in how to be a family person. In retirement, and from observing other families, I have had lots of time to think about how I should have done things differently.

The Sea's Enthrall: Memoirs of an Oceanographer

Anne and I returned from Paris with our two daughters, in 1964. In 1965, our son, Peter Gordon, was born. He was a great delight to me and the rest of the family. He was strong in body, and adapted well to life with two older sisters to boss him around. We should have had marital bliss, with three children and no financial problems, but that was not to be. Sometime during a visit to the United Kingdom, in 1975, I noticed that my eldest daughter, who was a very pretty blonde, was getting thinner and thinner. Actually, her decline was so rapid that one day, I failed to recognize her in a swim suit on the beach. We started to talk about her tendency to diet and I said that she was overdoing it. Her response was a complete denial.

"Leave me alone," she said, "I know what I am doing. It's my body."

Little did I know that this was just the start of a very long battle and one that, in spite of every effort, we were destined to lose. I could not get her sudden change of appearance out of my mind, and I did not know what was wrong, except that she had developed an iron will against eating.

On my way home, ahead of the family that year, I read an article in *Time* magazine about the "Twiggy" syndrome—people who starve themselves, otherwise known as anorexics. We contacted our family physician as soon as all the family were home and he confirmed our worst suspicions. Stephanie's decline was so rapid, that even with medical attention, including hospitalization, she died, in October, 1975, aged 15, weighing 67lbs. We were devastated. There is no more heart-rending an experience than to see a young and beautiful girl slowly destroy herself for reasons that neither she, nor the medical profession, understood. I had thought many times about protecting the family from crooks, illness, dangerous locations, and an assortment of other circumstances that all parents are aware of, just to have this insidious condition creep under the door of our house and totally destroy my eldest daughter. It was a terrible experience, and because of its peculiar nature, I am not sure, even today, if I was partly to blame for not understanding her condition several years in advance, which is what would have been necessary to offset it.

My family suffered from the loss, especially my wife, Anne, who devoted much attention to Stephanie during a time when it was very difficult to get her to see any reason. We felt betrayed by the doctors for not anticipating the rapid decline. Her attending physician never visited our house to study the home environment, which might have led to ways we could help her. Those treating anorexics today regard this an important part

Life in Home Port

of the approach. The same doctor showed his lack of compassion and concern by waiting six months after Stephanie's death before writing a letter of condolence. In the meantime, he wrote a scientific paper about her that appeared to me to be a cover-up for his own inability to prevent her demise.

I wrote a short book about her condition, based on her own writings. The theme of the book was published by *Readers' Digest* in the October, 1977 edition. This fulfilled one desire in her diaries, which was that her story should be known by many people. *Readers' Digest* claimed to have a circulation of 14 million at the time. Stephanie loved to draw cartoons and write poetry, during her short life. Here is one of her poems, which reflects her unhappiness about moving from the country to the city, in 1972. It reminded me of my much earlier experience of moving to Montreal.

City Life

I hate the crowded city street
The hot stuffy bus
The noise of squeaky feet
And people that fuss
On a hot August day ...
... in a town.

Life continued, but things were never really the same. As sometimes happens after such a calamity, relationships came under a strain, which affected the family dynamic. Three years later, my marriage to Anne ended; another calamity, and although we are now the best of friends in other ways, the business of living together was not a great success. We went our separate ways in 1978; a sad, but not bitter, parting. I was certainly to blame for the breakup and it is not something that I am proud of, except to say that this just seemed to be part of living. I tried to keep up with Allison and Peter, my two remaining children, but this was not wholly successful.

I continued my work at the university, but in 1979, took advantage of an opportunity to do some research in Australia. At that time, I came to know a younger colleague from Canada who was visiting "down under." We developed a close relationship. She was a marine biologist and so we had something in common, with respect to careers. We had some wonderful times, but our age difference was not healthy for the long term, and we parted.

Later, I met Dr. Carol Lalli, who was a professor at McGill University. She was a scientist of considerable note in the field of certain planktonic organisms, called pteropods (winged angels), a species of snail that becomes superabundant at times, and is a major source of food for some whales. We had a lot in common. She was very beautiful, soft spoken, and intelligent. Her conversation was lively and she was a gourmet cook. She joined me in Vancouver in 1979, and we were married in 1980. We set about writing a new textbook, which has now been widely circulated and translated into Japanese and Chinese. We traveled together a great deal: to China in 1982 for a series of lectures, later to Taiwan, the United Kingdom for six months in 1986, and to Japan in 1990, where I was appointed Distinguished Foreign Professor for three months. Carol has also conducted her own research on projects in Western Australia and Thailand. We bought a townhouse in Kitsilano, which is an interesting part of Vancouver, composed of small shops and restaurants. It was also close to the university, where Carol was given an honorary appointment.

Carol's specialty in marine research is in the field of mollusk research and, in particular, she studies a group of free swimming snails known as pteropods (meaning "winged angels"). Most of the world's mollusks live in the sea, and they range in size and complexity from the giant squid, the octopus, all the clams and snails, down to some minute swimming snails. Divergent lifestyles include the giant carnivorous squid (30m or more in total length), the sedentary snails that browse the seaweeds or filter phytoplankton, carnivorous snails that fire poisoned darts to capture passing fish, and those smaller mollusks belonging to the zooplankton community. These last may be carnivorous, feeding on other swimming snails, or herbivorous, feeding on the phytoplankton. This group includes the pteropods, and they are sometimes so abundant that they form the principal food of the baleen whales. My wife is one of the few experts in the world on this fascinating group of organisms. In her lectures she often tells the story of how some members of this group are totally homosexual in their reproduction. Each individual first becomes a male, and mates with another male. When this excitement is over, these individuals grow and become females. Their eggs mature, and then individuals use the sperm stored from the earlier mating for fertilization. This reproductive strategy teaches us that gay sex is not wholly an abnormality in Nature.

Some time after our return from China, I came down with meningitis.

Life in Home Port

It took a while for the medical profession to diagnose my almost perpetual headache, which was soon accompanied by vomiting and other nasty symptoms. The most telling diagnosis came from a spinal tap, which showed far too many white blood cells in my spinal fluid. Fortunately, further diagnosis showed that it was viral meningitis, which is the one form of the infection that is much more curable than the bacterial or fungal types. For three months, I was greatly weakened by the disease. I lost 35lbs and looked as if I was on the way out. A friend's wife phoned me one day, anxious to know if I had any permanent brain damage. I said that I could not tell because I had not used my brain much lately. A potentially life-threatening disease is a sobering soul searching experience. However, I was lucky, and was back to teaching, research, and of course, tennis, within about four months.

My son was not so fortunate. He died in 1998. He had been ill for some time with ulcerative colitis, but this had subsided for at least five years after he had a colonectomy. He was doing really well, when he suddenly succumbed to cancer that had followed on from the intensive hormone treatment prior to his major surgery. Much earlier in his life, I had resisted his having the colonectomy, and it was his decision later on. I realize now that it should have been done earlier, but hindsight is always more illuminating. His death was another terrible experience, as anyone knows who has ever watched a cancer patient slowly drain away, often in agony. He was married at the time, and had two very lovely children.

Earlier in his life, he did not seem to have had any idea what he wanted to do. After about five years of low paying jobs, he suddenly discovered that he liked accounting. He became the chief accountant for *Young Life*, an organization devoted to helping children. By all reports, he was an excellent accountant, which was a coincidence because my father had been an accountant, but whatever the genes are that make accountants, they certainly skipped my generation. Another coincidence was that both my father and my son died at the same age, and both left a widow and two small children.

With the deaths of two children in my family, I started to write a diary on a more or less regular basis. Apparently, this is a common reaction to the loss of family, and I do not regret trying to put down some of my inner feelings and examining them more closely. In a diary devoted to the understanding of one's inner self, *Marcus Aurelius Antoninus to Himself,* the famous Roman describes his mission to "face the facts" in a "world that is one living whole, of which each man is a living part." Marcus Aurelius lost four of his five sons, and Stoicism was his creed. As part of the living whole, what happens must be right.

"All that befalls an individual is for the good of the whole."

This abstract view certainly removes one from the immediate scene of personal tragedy. Scientific success was one thing, but it had taken a terrible toll on my family, and if one could not dwell on the bigger picture of life, then it almost appeared that family tragedies were some kind of penalty that had to be exacted.

My diary entries were mostly about thoughts and not about events. They were a kind of therapy that I found useful in releasing my inner feelings. I started to seriously question the basis for my religious beliefs, and shall write further about these feelings in my discussions with a Russian immigrant, Alexander Vishnevsky. I found it challenging to comment on all the books that I was reading, and I became interested in poetry.

All this reading and writing was a way to cope with the transition from one life to another. I had moved from a family life with Anne, to another life-style with Carol. My life had been greatly enriched through Anne, and as time progressed, through Carol. Each, in their way, has contributed to helping me, the one with family, and the other with my profession. My remaining daughter, Allison, took a doctorate in psychology at the University of British Columbia. She has one child, who was born halfway through her final examinations. I was never sure which came first, the doctorate or the child by Caesarean. I have been very proud of her academic and family successes, the latter being, in part, due to a wonderful husband, Wayne.

Having been born to richer parents than I was, she does not seem to have inherited some of my peculiarly frugal habits. For example, I have cut my own hair for most of my life, since it was a saving during college days, while on a meager budget. Also, I love to darn my own socks and sew on buttons. I think that I get these habits from my grandfather, who was almost self-contained in his domestic care. He even cobbled his own shoes, and was known to have pulled one of his own teeth. New generations have new ways, which is all part of progress.

Apart from family life and my work at the university, I had another major interest during this period in Vancouver. I have always enjoyed meeting people, and so I decided to become a volunteer for senior citizens, with the Vancouver Health Department. I wanted to help people who needed companionship or mental stimulation, since I believed friendship and having someone to talk to about thoughts and feelings were important for anyone, especially the aged. I was also interested, on my own account, in talking about the deeper issues of life with persons who had experienced as much as, or more, than me.

Over a ten year period of visits, my first and most memorable contact was just the person for such explorations, and I know he added as much, if not more, to my life than I did to his. Alexander Vishnevsky was a Russian who had escaped during the revolution under difficult circumstances, when the communists took over Russia. He had been born around 1900, to Jewish parents in the town of Nikoli, in the Crimea.

"When I came to Canada in the 1940s," he told me, "I changed my name to Hugh Winslow. I felt it was necessary, in order to avoid any discrimination."

After working with a Doukhobor newspaper in the interior of British Columbia, he contracted TB in 1952, and spent four years in a sanatorium. His life changed dramatically when he awoke from major surgery on his lungs, hearing music and voices. He was officially diagnosed as schizophrenic. The Provincial Mental Hospital in Essondale became his new home.

Hugh, however, was not content to stay, and he escaped. There followed years of poverty and dependence on a few trusted friends. He lived under bridges, scraping by on a diet which consisted mainly of sardines, but he was never recaptured. Legally, after five years of freedom, a person was deemed sane enough to be independent of institutional help, but Hugh never asked a doctor for another opinion. He was afraid of being sent back to Essondale.

"I still hear voices," he said, "but they don't bother me like they used to."

When I met Hugh Winslow, he lived in a small apartment in Vancouver. He was a frail figure, and his voice was rather weak, due to little lung capacity. I noticed his exceptionally good manners and his impeccable dress; this, in spite of his extreme poverty. His face was very thin, his hair well brushed, and he wore colored shirts, always with a tie, which was usually brown or black.

Hugh had his own way of dealing with his lack of money. He had surrounded the chair behind his desk with a large cardboard box, into which he could climb and sit down, as if he were in a sedan chair. Under this chair he had placed a 100W light bulb, which produced enough heat in this confined space to satisfy his bodily requirements without extra heating of his apartment.

His ingenuity did not stretch to dealing with cockroaches. Although his subsidized apartment was almost new, it was infested with these creatures and Hugh had to keep all his food wrapped in plastic bags. Despite his precautions, cockroaches appeared many times when I visited him.

These distractions did not affect our relationship. One of Hugh's main preoccupations was with the purpose of life: since we live in an evolving world, he argued, there must be purpose to our evolution. I was very willing to discuss this with him. It was an unusual role for a volunteer health worker to help exercise someone's mind instead of taking him for a walk, but I was intrigued by the challenge. As we talked, his ideas began to influence my own thoughts in ways that I would never have anticipated.

He began by acknowledging that while people are very ready to debate objective issues, they are far less willing to discuss more personal topics such as their own beliefs. He, on the other hand, was very eager to embark upon the big questions, the meaning of life and death, religion, the nature of humanity. These were subjects of great interest to me also, and I was fascinated by his perspective. Hugh believed that the universe was created from a restless void that expanded—in the same way that zero can be infinitely expanded mathematically—into positive and negative terms, as long as there is a balance of each. This allowed for the concept of opposites, and the connectedness of all events in the universe, however remote.

"There is a balance in life between love and hate, up and down, light and dark. In fact, there is dualism in everything we know." He referred to the Chinese philosophy of the Yin and the Yang, and I agreed with him, citing passages from the Gnostic Gospel of St. Thomas.

"Yes," I said, as I was reminded also of a statement by the famous physicist, Niels Bohr, who said, "The opposite of a profound truth may well be another profound truth."

We were interrupted when a large cockroach started to wander across the top of Hugh's "sedan" chair.

"We have a visitor, Hugh. Where's the swatter?" I asked. I found it on his desk and aimed a hefty blow at the cardboard above his head. It must have sounded like an explosion inside his cubicle. He asked me to be a little gentler next time.

"Well," I continued, "why should the universe evolve towards greater and greater complexity once it is created?"

He replied that it was not a contradiction of the second law of thermodynamics to say that in an open system, given an almost infinite amount of energy, a state of organization can be extracted from that energy. One expression of this is that the energy of our sun is equivalent to the release of 10^{38} bits of information per second. He also gave pragmatic examples of the large amounts of energy needed to build a microscope or a telescope, and the corresponding, enormous amount of information to which they gave access.

He went on: "The increasing efficiency of energy use is a necessary part of the evolution of human society. Our most evolved societies, such as the USA, Europe, and Japan, are the ones that have the highest per capita consumption of energy, and are also the most informed. The infinite possibilities of acquired information represents, at least to our present state, a godlike condition, just as it seems almost incomprehensible that our present state has evolved from unicellular organisms over time."

"So you are saying that our ascension from nothing, to an immortal state of consciousness, is the purpose of our being?" I asked.

"Yes," he agreed, "but how well we follow this destiny appears to be another question. The world has had many mystics who have pointed in the right direction, but one must be cautious of a too literal interpretation of mystical expressions. The same difference exists, by analogy, between the ancient alchemist who thought that he could turn other elements into gold, and a modern nuclear physicist who actually can. Much religious teaching is allegorical; in fact, nuclear physicists also use allegorical descriptions."

As we talked, I realized that our conversations emphasized my own search for some new perspective on religion. Writers and thinkers throughout the centuries have attempted to present new insights into how man sees his outer and inner universes, to reflect on this relationship with the material, and the possibilities of a spiritual, world. Many of them, however, such as the different orthodox religions, base their beliefs on ancient, revered texts, and are unable to reach a consensus. Modern science has added new dimensions to the debates, with the exploration of the genetic base of life, and new knowledge about the outer reaches of the universe.

My discussions with Hugh inspired me to continue a search for truth, a search that goes on and is unlikely to end. Since associating with him, I have come to explore more earnestly the works of thinkers such as Teilhard de Chardin, Julian Huxley, Albert Einstein, and Julian Jaynes. I have explored the Gnostic gospels, particularly interpretations given by Elaine Pagels.

In *The Gnostic Gospels*, Elaine Pagels looks at the question of how the Christian religion evolved in the first place. Some of the early Christians were known as Gnostics, and they anticipated that the present and future would yield a continual increase in knowledge. They came into competition with Rome, whose authority on the interpretation of religion became increasingly powerful, in a material sense. The Gnostics had no such authority with which to protect themselves. They were declared heretics by the Church of Rome, and brutally suppressed. These suppressions eventually included their own adherents, such as Bruno and Galileo, who sought truth through observation. This history of warring factions within Christianity continued from the beginning and continues today. It is not unique to Christianity, however, and is generally found throughout the world's religions.

The problem of choosing between enlightenment and traditional religion is aptly described by Shakespeare in *The Comedy of Errors*:

> *Let us once lose our oaths to find ourselves,*
> *Or else lose ourselves to keep our oaths.*
> *It is religion to be thus foresworn;*

It is a common adage of those who promote absolute faith that "too much analysis leads to paralysis." This is, perhaps, where the pathways of the scientist and the theologian diverge. It has been said that when the evidence conflicts with the theory, the scientist rejects the theory. The theologian rejects the evidence. Science does not specifically incorporate a deity, yet to many scientists the wonderful intricacies of Nature can inspire religious curiosity This is not based on archaic texts, but on the gradual revelation of all reality through science. To fundamentalists, such a religion would appear to lack a moral order. However, it is the teaching and practice of what is claimed to be a moral order that has led to so much conflict and persecution in the history of traditional religions. Moral order could be attained through the courts for everyone's benefit. On the other hand, enlightenment can only be achieved through heightened awareness. In the words of Julian Huxley (*Religion without Revelation*):

> *Man is that part of reality in which and through which the cosmic process has become conscious and has begun to comprehend itself. His supreme task is to increase that conscious comprehension and to apply it as fully as possible to guide the course of events.*

With this in mind, religion becomes more of a quest than a blind belief in ancient texts.

Among today's religious leaders, I find that the need for change in Christian teaching has been well expressed by Bishop John Spong, an Episcopalian who has joined others in abandoning the father-figure kind of god, searching, instead, for a presence more relevant to everyday life and the knowledge gained throughout human history. In a new religion, we must accept responsibility for our own lives and not abdicate it to an ancient concept of a protector god.

So, from my upbringing as a strict Anglo-Catholic, I became exposed to many powerful thoughts of the past century. This was, in part, due to the conversations I had with Hugh Winslow, who introduced topics I had not thought about since my time as a college graduate. He reopened a world of ideas for me and, hopefully, I gave him the mental comfort of having someone to listen to his own interpretations of life, before he died, in 1978.

X
The Japan Prize

1. The origin of The Prize

    The Japan Prize is bestowed by the Science and Technology Foundation of Japan (STFJ). It was established in 1982, as a prestigious international award in the fields of science and technology, and was motivated by a desire to express Japan's gratitude to the international society for the national prosperity of Japan. The STFJ is an independent organization sponsored by government and industry. It was formed to take over the idea, proposed by Mr. Konosuke Matsushita, that international prizes should be given in support of "peace and prosperity."

    Original funding came from the late Mr. Konosuke Matsushita (1894-1989), who was the founder of the Matsushita Electric Company. He was born of poor parents, who both died before he was sixteen. At the age of nine, he went to work in an Hibachi factory, but one year later, he was working in a bicycle shop. At the age of sixteen, he went to work for Osaka Electric, where he was put in charge of teams installing electricity in private houses. At twenty-two, he founded his own electric company and started to sell an improved light socket. He developed the idea that everything could be made better and cheaper. He followed this idea in developing his own bicycle lamp, radios, and a variety of other electrical products. He started the trade names, National and Panasonic. His companies were often called "copy cat" companies because he invariably improved on designs, rather than inventing new products.

The Panasonic slogan "Slightly ahead of its time," is perhaps a more fitting description of Matsushita's philosophy. The Sony company, in contrast, was considered to be on the cutting edge of technology. Mr. Matsushita had a religious experience in 1932, and from this evolved his idea that industry must not be used for personal wealth, but to eliminate poverty. He was featured on the cover of *Time* magazine in 1962, and by 1984, he was the richest industrialist in the world, from his ownership of shares in the Matsushita Electric Company. At that time, revenues of the company were just under 50 billion dollars per year, compared to the Ford Motor Company revenues of 10 billion, and Microsoft of 4 billion. He was described as a rather shy and humble man, by those who knew him. He was married, with two daughters, but he also had a mistress, which was normal for Japanese society. He was the benefactor to a great many national and international foundations in Japan, North America, and Europe.

The Japan Prize is open to all nationalities. Two awards are made each year, in different categories, which are predetermined several years in advance. Nominations are by individuals or groups, and selection of a winner is made by the STFJ. Each laureate receives a certificate of merit, a commemorative gold medal, and a cash award of Y50 million (approx. US$500,000). The Prize has sometimes been referred to as being equivalent to a Nobel Prize in science. Its value is similar, and several Japan Prize laureates have won Nobel Prizes, either before or after the award of a Japan Prize. However, they differ in the fact that the Nobel Prizes are always awarded in fixed categories (e.g. chemistry, medicine, etc.), while The Japan Prize categories change every year and include technology, as well as science.

2. The award

My wife and I were in New Zealand in November, and came back, toward the end of the month, to the usual assortment of correspondence and all the catching-up that is necessary after a long trip. Unknown to us, a fax was waiting at the Institute of Ocean Sciences, where we both have honorary positions. I went to play tennis the second morning after our return, and my wife took a phone call from the office of the Consul General of Japan. The caller asked what our reaction was to the faxed information. In anticipation of something important, my wife immediately drove up to the office and picked up a fax announcing that I had been selected as one of two 17th laureates for The Japan Prize. She brought this information to me on the tennis court, where my fellow players were rather indignant at the interruption of our game. My first thought was "Wow!" I wasn't sure whether I should stop playing and start celebrating, or continue the game. My tennis buddies decided that they wanted to finish the match, so with wild abandon I started to swipe the ball in all sorts of directions, until I was allowed to leave the courts.

My Japan Prize had been awarded in the field of Marine Biology and the other laureate, Dr. John Goodenough, an American from the University of Texas, was given his award in the field of Science and Technology of Environment Conscious Materials. He discovered the lithium cobalt oxide battery, which is now an indispensable power source for mobile communications, and will likely become a power source for hybrid and electrical cars. At the time of his discovery, Dr. Goodenough was chairman of the Chemistry Department at the University of Oxford, England. He told the story, in Japan, of how he first took his idea to some electronic companies in the United Kingdom, but how they showed no interest in the commercial potential of his discovery. He was later invited to lecture in Japan, where he found an immediate market among Japanese manufacturers for his product. It is surprising how much industrial enterprise has been lost by Britain during the last fifty years, due to a lack of such awareness, and how often this has been picked up by Japanese manufacturers.

At the time of my birth, and through to some years after the war, Britain was a leader in small car manufacture, motorbikes, shipping, and telecommunications. By ignoring certain crucial innovations, such as the replacement of the kick-start motorbike with a battery-started bike, Britain managed to fall behind in many industrial areas, where they had formerly led the world.

Long citations giving the reasons for each award accompanied the announcements. A précis of my citation was:

> *Through his research devoted to obtaining a holistic understanding of how pelagic organisms are interconnected in the trophodynamic food-web of the sea, he has made a great contribution to the development of Biological Oceanography. His goal has been to present an alternative method for the management of fisheries, based on measuring the dynamic relationship between fish and their physical, chemical and biological environments.*

When I read the citation, I realized that, of course, many other scientists had been working towards the same ends (I have already mentioned some of these), together with others who seemed to be working toward a different scientific approach. Citations such as the above are built on the shoulders of other scientists, and both Dr. Goodenough and I acknowledged the work of colleagues in our acceptance speeches.

The University of British Columbia and the Department of Fisheries and Oceans, where I had begun my career, were the first to celebrate the award. Dr. Martha Piper, President of UBC, arranged a special press conference, which was attended by the Japanese Consul, Mr. Kusumoto, the Director of the Department of Fisheries and Oceans (Pacific), the Deans of Science and Graduate Studies, the Department Chairmen of Earth and Ocean Sciences and Zoology, where I formerly held appointments, and by a number of my colleagues in oceanography, whom I was particularly pleased to see. Another news conference was held later, at the Institute of Ocean Sciences, near Victoria, BC, hosted by Mr. Robin Brown, Acting Director of the Institute.

In Japan, there is a whole week of celebrations connected with The Japan Prize. This is held in Tokyo, and includes visits to the Academy of Sciences, to the JFST, to the Prime Minister, and finally, a presentation ceremony and banquet attended by the Emperor and Empress of Japan. In addition to these meetings, receptions were held at the American and Canadian Embassies, and Dr. Martha Piper also held a special 'UBC night' reception. This was a very busy schedule, but somehow, the whole procedure went off with clockwork precision, including a final visit to Kyoto, where my wife and I were entertained by the Matsushita Electric Company, and dined with Konosuke Matsushita's grandson, Mr. Masayuki Matsushita.

3. The personal side of our visit to Japan

Between the announcement of The Prize in November and our taking part in The Japan Prize Week celebrations in April, there were many preparations to be made. First, prior to my arrival in Tokyo, I was given a time schedule for the submission of speeches, which included my Commemorative Lecture and my Acceptance Speech. There were other speeches to be given at a press conference, at a meeting of scientists, at the banquet, and during our visits to the Academy and the JSFT. Of these speeches, the Commemorative Lecture was the longest, and needed the most preparation, in terms of both the talk and the illustrative material. I decided against a PowerPoint presentation because of the difficulty of matching computers, and the embarrassment of trying to correct any malfunctioning equipment. I relied on about thirty slides to take me through the talk, and I noted that my colleague recipient safeguarded himself further by only using an overhead projector. With large audiences and an important lecture, one does not want to take unnecessary risks.

In addition to work on the preparation of these speeches, my wife had to have some formal dresses and, in the end, selected three dresses, one white, one blue, and one black, which I felt was at least one too many, but as it turned out, she had occasion to use all three. My own formal attire had to be rented for two weeks, and I had a new dark suit made for the less formal meetings. Travel to Japan was by First Class accommodation on Japan Airlines but, unfortunately, no airline was offering such accommodation on the days that we were booked, so we ended up in Business Class.

On arrival in Japan, we were met by the executive secretary of the JFST, Mr. Morihisa Kaneko, who whisked us away in a black limousine, our luggage following by taxi. At the Hotel New Otani, the manager met us with a number of his staff, in formal attire. We were personally conducted by him to our suite. This consisted of three rooms, beautifully decorated with several *ikibana* arrangements of flowers, and with a magnum of champagne on the coffee table. The manager assigned us one of his assistant managers to call personally, if we had any requests and, in fact, this person was always waiting at our door in the morning, to conduct us downstairs for the first event of the day.

On the first day, we had a briefing session in which the whole procedure of The Japan Prize Week was described to us, and a private dinner that evening was arranged with the president of the Foundation, Dr. Masami Ito, and its chairman, Dr. Jiro Kondo. These were two of the most distinguished Japanese that I have met, and their dinner-time conversation in the English language was lively and entertaining. On the second day, we were introduced to senior members of the Japan Academy, and a press conference was held in the afternoon. After some polite questioning, one reporter homed in on the Japanese whale harvest and asked me what I thought about it.

This is an issue which has made Japan unpopular in the world, but one which seems to have been attacked on two fronts. The first is that there are persons who are emotionally charged when it comes to the killing of whales in any circumstances, deeming it to be a disgrace for such a noble beast to die in such a way, unnecessarily. It would seem a fair argument, except that it should be equally applied to all animal life, which means we should all become vegetarians. This would certainly solve many global problems, and may eventually occur, but at present, it is too soon in the evolution of human society. The other part of this problem is whether or not there are species of whales that are endangered. The minke whale has recently become superabundant and it is the species most often targeted by the Japanese whale fishery. Its abundance may, in fact, be preventing the resurgence of the blue whale population, since the latter feed on the same food as the minkes. On this basis, some harvest might be allowed of minkes. I am not a whale expert, and I suppose that was clear in my answer to the Japanese reporter, who said in his article (translated) that the professor "seems to have mixed feelings on this issue." It would have been interesting to ask Alfred Nobel what he thought about blowing up a number of workmen in his explosives factory, when he was still experimenting with dynamite, to say nothing of all those who died in the First World War. If a recipient of The Japan Prize for Marine Biology could become a target for comments on the the whaling issue, why were winners of the Nobel Prize for Chemistry not badgered about the evils of TNT?

The next event that we were supposed to participate in, the visit to the Prime Minister, had to be cancelled, as it was the very day the incumbent minister and the new were exchanging office; it was reasonable that this did not allow much time for a visit from us. On Thursday morning, we went to check arrangements for our Commemorative Lectures, and in the afternoon, Dr. Goodenough and I presented our talks to a mixed audience of about 400 scientists and laymen. I must say that it turned out to be a lot easier to make marine biology sound interesting, than to listen to the finer points of the design of Dr. Goodenough's battery. No slight intended on his excellent presentation or masterful work, but his technical expressions were a little overpowering, while I had an advantage of being able to show beautiful pictures of jellyfish and other marine creatures.

In the evening of our second and third days in Tokyo, we were entertained, first by a reception at the American Embassy, and then, the following evening, at the Canadian Embassy. Both receptions were most enjoyable, in that they gave us an opportunity to meet many of the persons behind the scenes in marine affairs in Japan. The Canadian Ambassador, Mr. Leonard Edwards, gave a reception and spoke eloquently about the importance of The Prize. As it turned out, by sheer coincidence, it was also "Think Canada Week" in Japan—an occasion in which it was hoped to advance ideas about Canadian technology. Naturally, the occasion of The Prize fitted very well into this context. Mr. Edward's embassy reception was about twice the size of the American Embassy reception, and it seemed that many ambassadors of other countries were present. I have never been on a reception line in which I was required to shake hands with so many guests. It was amusing to find that while the Ambassador was in complete control, the chief chef became very agitated at some point with the slowness of our reception line-up. He passed the word to his Excellency that we should pick up the pace a little!

During our stay in Japan, and both before and after our departure, the scientific attaché to the embassy, Dr. Philip Hicks, was most helpful to us in a variety of ways, including helping with documents and transportation, making contacts, and providing translations of the Japanese press accounts of our meetings. Compared with my experience visiting embassies in Africa while with UNESCO, this was one embassy that seemed to function well.

On Friday, we started the day by going to a rehearsal for the presentation ceremony that was to be held in the afternoon. Morrie Kaneko, the Executive Secretary of JFST, was extremely precise in his instructions because anything involving the attendance of Their Majesties required a flawless performance. When to bow, who to bow to first (the Emperor, not the audience), where to sit, and the order of sitting down, how to receive The Prize, and the precise length of the acceptance speeches were all duly practised. It had the required effect, and the afternoon performance went off like clockwork. The last to arrive before the ceremony were the Emperor and the Empress. Before we entered the theatre, they greeted us with a friendly aura of regal simplicity, almost as if we were now 'family'. The royal couple then went ahead into the theatre, where everyone was seated. We two laureates and our wives, in formal attire, marched onto a stage magnificently decorated with flowering cherry blossom trees that had been especially cultivated for the occasion.

The theatre was filled with a thousand guests. We traversed a raised walkway to the music of "Land of Hope and Glory," and the audience, dignitaries, and Their Majesties all clapped as we took our seats. The Tokyo Symphony Orchestra played the national anthem, and the presentations were made by the President of the JFST. The Emperor concluded the speeches with a short speech about the work of the two laureates and the importance of science in the world.

After the presentations, we all moved off the stage to the front row of the balcony, where we sat on either side of Their Majesties, to listen to a concert given by the Tokyo Symphony Orchestra.

The choice of music had previously been made by the two laureates. My first choice was for the overture to *Coppelia*. I like this ballet because it is about a professor who starts out trying to do one thing, fails, but then discovers he has done something infinitely better (by having two people fall in love as a result of his efforts). A second choice was Bach's Orchestral Suite No.2, which has a flute solo, and was especially favored by my wife, who plays the flute.

After the concert, we had a couple of hours before the banquet. Carol changed her dress for the banquet, which was also a formal occasion. The menu was entirely French cuisine, consisting of a vegetable mousse, ginger flavored fish, *granité aux pétales de rose*, a medallion of beef, caramel ice cream, and coffee, with appropriate French wines. The Emperor proposed a champagne toast to the two laureates. During the banquet, an eighty-person choir sang two folk songs from each of the recipients' countries, as well as a selection of Japanese melodies. For the United States, the national choice was "Beautiful Dreamer" and for Canada—that was more difficult! How many folk songs are truly Canadian? I think that the choice came down to "Alouette" or "Farewell Nova Scotia." The latter was the choice and the Empress, who was sitting next to me, assured me that she much enjoyed it.

The Sea's Enthrall: Memoirs of an Oceanographer

During dinner, Carol was sitting next to the Emperor, who is a fish biologist, and so they had some kind of technical discussion. He also recalled our previous visit in 1981, when we first met His Majesty. At that time, I was invited to give a lecture to his father, the late Emperor Hirohito, in the Imperial Palace. Afterwards, we had gone to meet the present Emperor, who was Crown Prince at the time.

I sat next to the Empress, who speaks excellent English. We discussed childhood memories (she has written a book on her own childhood), and the game of tennis, as both of us are very keen players. On my other side, at dinner, was the Chief Justice of Japan. He did not speak very much English, so there was a lady translator sitting on the floor between us allowing us to communicate. After dinner, there was a special meeting for Their Majesties to become acquainted with young Japanese scientists. This was held in a secluded room, and it was a great opportunity to talk informally with the Emperor and Empress.

The next day, we had scientific discussions with a group of about thirty senior scientists in our respective fields. The persons selected for my discussion were mostly scientists with whom I had had a long acquaintance. We talked about ways to revitalize fisheries science, and not all the participants necessarily agreed with my ideas.

At about noon, we left for the railway station, where we boarded the Nozomi bullet train for a two-hour ride to Kyoto. We stayed at the Miyako Hotel in Kyoto, as the guests of the Matsushita Electric Company. The grandson of the founder of this company, Mr. Masayuki Matsushita, was our host during our stay in Kyoto. He arranged for us to visit the Golden Pavilion in the Rokuon-ji temple. We had seen this before, but visitors are not normally allowed inside. This time it was different and we had a conducted tour by the chief monk of this beautiful shrine. Mr. Matsushita also took us on a personal tour of the Matsushita Museum Gardens, where his grandfather had spent much time in contemplation. The museum part of the tour gave us an insight to some of the most treasured handicrafts of woodwork, weaving, and printing in Japan.

In the evening of the next day, we were entertained at the Tsuruya Restaurant in Kyoto, which is said to be the best Japanese restaurant in all of Japan. Our dinner host and hostess were Mr. Matsushita and his charming wife, who wore a very fashionable Izzy Miyake dress. We had a twelve-course Japanese dinner, sitting on tatami mats—actually they allowed us westerners the comfort of slightly elevated floor chairs with backs and a place for our long legs under the table. At the dinner, young geisha girls, known as *maiko*, were in attendance and joined in, at least with the drinking part of the meal. (The *maiko* lady is unique to Kyoto, and is below the rank of a mature geisha, who is known as a *geiko*.) They carried on a lively conversation with our host and charming hostess, and asked much about us. Our host was given a light back massage by a *maiko*, during part of the dinner, and I tried a couple of the very few of my Japanese expressions with one of them. These turned out to be funnier for how I said them, rather than for what I said. In the middle of the meal, the wall on one side of our dining room was withdrawn, electronically, to reveal a stage. The geishas then carried out some musical entertainment, which was quite delightful.

From Kyoto, we went back by Nozomi express to Tokyo, where we were met by Professor Seki. Our official Japan Prize Week had ended, but I was to receive an honorary doctorate from the University of Tsukuba. We were entertained by the President of the University, Dr. Yasuo Kitahara, who arranged for a special luncheon, followed by the conferring of the honorary degree. I gave a lecture to the audience of about 300. The subject, this time, was on the role of the oceanographer in understanding the problems of coastal pollution. We left the next day for Vancouver, and had a hard time readjusting to all the normal things in life, after being so magnificently entertained in Japan.

A few months later, I was again honored in Japan by the conferring of an honorary doctorate at the University of Hokkaido, in Sapporo. In a dignified ceremony arranged entirely for the single purpose of the award, the President and deans attended the event, along with an academic audience of about fifty persons. Afterwards, they hosted a delightful lunch, at which we were accompanied by two of my Japanese friends, Professors Ikeda and Maita. Professor Ezura, Dean of Graduate Studies, further contrived an additional ceremony of recognition in Hakodate, where I gave a public lecture and visited Deputy Mayor Mitsuya, in the city hall.

In the following year, I was surprised by the award of an honorary doctorate from the University of British Columbia. At the award ceremony, I was gratified to see that many of my colleagues in oceanography had joined the academic procession. I gave a short speech to the graduating class in science on the value of anomalous results in scientific data.

XI
A Refit for the Future

I had retired from the University of British Columbia in 1992, nine years before I was honored with The Japan Prize. I took early retirement and did not, in fact, have to leave for another five years. However, while I had benefited greatly from university life and facilities, I had become disillusioned with the university administration. Fund raising and economic organization were the primary concerns, and the development of science was a secondary consideration. I gave up on ever making any administrative progress in fisheries oceanography on the campus, and I saw more interesting opportunities as a retired person operating independently in ocean science. Coincidentally, very soon after I had made my decision to leave UBC, two other senior professors in our department also took early retirement.

Retirement was not a financial burden because the pension system had worked reasonably well. Conversely, there were many cases of professors staying to the very end of their appointments, and even a couple of professors who sued the university because they claimed it was age discrimination to let them go at sixty-five. Another professor had a mild heart attack at the age of sixty, recovered and then was able to stay on for five years while someone else did practically all of his teaching, and he collected his $100,000 salary, mostly at home. I believe that the university is, at times, unnecessarily generous to its loyal supporters. The tendency for a number of departments to allow professors to work off-campus during much of the year is another questionable practice. However, I believe that life has changed a great deal since a new President, Dr. Martha Piper, took on the unenviable task of managing the educational system for around 35,000 full and part-time students.

Having unexpectedly received a large amount of money as a result of The Japan Prize, I was now faced with the question to which some people seemed to want an answer. What was I going to do with the cash? The easy answer was to say that it was nobody's business but mine. However, having seen the whole procedure in my life leading up to this Prize as something which was as much circumstance as effort, I have felt that I owed my friends some explanation.

In the distribution of the money, I wanted my family to come first. My ex-wife, Anne; my wife, Carol; my surviving daughter, Allison; my daughter-in-law, Tracey, who had lost a husband as I had a son, and the memory of Stephanie, all prompted me to remember these persons as people who had shared, most closely, in my career, largely from the domestic side. I was glad to be able to share some of the money with them, as well as with organizations which affected their lives, such as Young Life and research on colitis at UBC.

Next, I owed a debt of gratitude to my school in England, which is a charitable foundation. Even though it has now become the richest school in the country, it still depends on charity for the support of students, many of whom may not pay anything for their education, depending on their family's circumstances. On leaving this school, all the boys and girls are given a Bible and the following charge:

> *Never forget the benefits which you have received in this place and, in time to come, according to your means, do all that you can to enable others to enjoy the same advantage; and remember that you carry with you, wherever you go, the good name of Christ's Hospital.*

I made a presentation to the school, to cover their educational requirements for a student, and an additional donation for a building fund.

Then, I considered that some of the oceanographic societies with which

I had been associated could use some extra finances for publications or meetings, and so I donated to several of these. At the University of Concepcion in Chile, I had participated several times in teaching programs, and once in reviewing the university's teaching objectives. We know very little about the Southern Ocean because it is of interest to relatively few countries. The University of Concepcion has taken a lead in training more oceanographers in the southern hemisphere, and I was pleased to be able to support these efforts in a small way. Also, I have some domestic charities that I have been giving small amounts to over the years, and I decided to gradually increase my contribution to these organizations.

There is a rather irritable point about such giving, however, and it is this. If, in one year, you give a large amount to a national charity, they thank you and then come back within six months suggesting that you might like to give more. This seems never to fail. In order to try and avoid this, I decided to increase my contributions per annum, but not in a way that would attract undue attention. I suppose that charitable organizations know that their tactics pay off, but I find the method disagreeable. It is difficult to know if my distribution of funds was the best way to deal with such an unexpected windfall, but it seemed to me to be a partial answer to the dilemma of how to remember others, while holding back from either a total retention of the monies or, as some have done, honoring a single charity.

When I planned to retire from the university, my wife asked me what on earth I intended to do. Like many wives, she did not want me sitting around the house all day, so her question was justified.

At the age of sixty, I had some ideas which I might call my "thirty-year plan" for the future. It did not, of course, anticipate receiving any honors, and so my plan did not take into account that such an event as The Japan Prize might open up new avenues of interest. In general, I have not followed up on many of the possibilities that The Prize presented, such as an increased load of public and academic lecturing, but I have enjoyed focusing on my memoirs, in order to reconstruct the many fortunate things that have happened in my life. I have done a little more speaking, to such organizations as Rotary Clubs, but except for this, I expect to continue with my "thirty-year plan" as originally devised, and I hope that I am not alone in thinking optimistically about the future.

Retirement reminds me of another time in life that I really enjoyed—my pre-school years. In retirement, one can go back to an age when you do not have many daily responsibilities. The analogy is taken further by some who refer to "men's toys," as expensive hobby or sports items that men finally find they can afford, often in their retirement years. One of my professor friends took this idea too far and planted an orchard of fifty fruit trees because he always wanted just that. However, he found that they needed constant pruning and other care, much of which involved climbing a ladder, which is a common hazard for older people. He had to give up on his orchard fairly soon after it was planted.

So far, my "retirement" years have been among my happiest in a lifetime. I shall work to keep them that way. The following are some thoughts on how this might be possible.

It is no medical secret that retirement is generally accompanied by a decrease in mental and physical activity. This may not be as detrimental as it first appears. A more relaxed approach to daily events leaves time for more humor. My much loved game of tennis now has an atmosphere of joviality that it never had before. There are the players who, in about the fourth game ask, "What's the score?" Opinions vary from 40-15 to15-40. After some discussion, an equable 30-30 is agreed upon, but then we have forgotten who was serving! Of course, we are not really forgetting this information; it is just that it is not very important any more. The game and the social contacts are what are important. Compared to competitive tennis, which I played at college, seniors' tennis is a totally different game. Line calls and aching muscles are a source of amusement and the social introductions to others in different walks of life are very rewarding.

Nevertheless, my number one concern is good health, which I have generally enjoyed. I want to stay as healthy as possible, and this means being physically active. I have retained my very active program of tennis and enjoy bicycling, walking, and boating. This is not enough, however, because as we enter our senior years, it seems that despite constant activiy, tendons and muscles continue to tighten, so that a regular program of stretching, along with some biochemical enhancement from vitamins is almost obligatory. Losing weight is also helpful, and I found this comparatively easy, as long as I rigorously followed a certain diet. My goal was to lose ten pounds in thirty days. This I did, on a diet of sardines, boiled eggs, raw vegetables, Ryvita, and condiments such as fat free mayonnaise, balsamic vinegar, and a delightful Japanese spice, called Sansho (or Japanese pepper). One or two 'dining outs' with full fare during this period did not affect the result. It was all a lot cheaper than buying a book or attending a course on dieting; in fact, I actually saved money!

Equally important as the health of my body is the health of my mind. During my career, I was often bothered by many events, including personal relationships, national issues, and international situations. I could not help this. I was that kind of person. Judges, for instance, who rule that it is okay to kill your wife, if you are too drunk to know what you are doing, give an open invitation to murder. Judges who wanted ancient hearsay evidence included in aboriginal land claims, have caused nothing but confusion in our society. Warring factions in Israel and Palestine, Northern Ireland, Spain, Sri Lanka, and the Balkans, all disturb me because potential solutions are always destroyed by bigotry, religion, or stupidity, while all around, innocent people are dying. Newspapers seldom report happy events.

Instead of being affected by all these events, it is much easier to stop reading about them. Unfortunately, they also turn up on television, where, in one year, there are about 500 news reports on Israel, but only twenty on Australia. I am so tired of hearing about the Israelis and would much rather hear about Australians! Removing areas of stress in one's thinking seems, to me, to be important. Replacing them with mental exercises, such as doing a crossword puzzle a day, learning another language (I am in my third year of Spanish lessons), learning to play a musical instrument, doing volunteer work, or taking a course in something you know nothing about, such as antiques and painting, in my case, are all a lot more fun than reading the news.

Another very satisfactory retirement occupation is to have a hobby in some area not immediately associated with one's work. I have always wanted to carve but never had much time. Using a supply of discarded whales' teeth that I acquired over many years while working in marine laboratories, and from the site of whaling stations, I have carved a significant number of whales, penguins, snakes and pendants, all of which I have enjoyed doing. Stamp collecting, left over from my boyhood days, has resurfaced as an attractive hobby. Drawing and painting are delightful subjects that we have little time for during our professional careers. I particularly like to draw cartoons, and I laugh at my own results, regardless of how others may see my humor.

If there is one mission left to me in society, it is to reduce the amount of noise pollution that surrounds us. I have been a member of the Right to Quiet Society (R2Q) for a number of years. Their mission has been to have ghetto blasters banned from parks and public transport, to reduce music in many dining places and, in general, to take up the problems of barking dogs, motorbikes, leaf blowers, and many other loud noises that surround us. I thoroughly commend their efforts.

It is said that in their first ten years of retirement, everyone travels, and in their second ten years, they stay home and look at all the photographs. Travel has become very easy for older persons and even many of the prices are quite affordable. Having traveled a great deal during my career, I am not obsessed with the need to travel more, although I think that one or two short trips over a year can be inspiring experiences, as long as one can stay away from places where there is armed conflict or dangerous health problems.

I think that I have dealt with most legal and financial matters in my life but, unfortunately, situations change, and one is never quite finished. It is very important, in both fields, to have friendly and honest advice, even if it costs a little more than expected.

Being independent to a degree is something that I value very highly for my state of mind. I never wish to have to depend on my family or complain to them about any of my conditions in life. I value the company of family, but I have also always valued being on my own, and this independence is something I feel is threatened as we get older.

There is a question in my mind as to how far one walks away from one's profession. At least for now, I have accepted honorary positions with the government and the university, and I have continued to take part in scientific teaching and writing to some extent. When contentious issues about the environment or the exploitation of ocean resources surface, however, I find that the situation becomes stressful, which imposes on my number one rule in retirement, to get rid of stress. However, one can not avoid following some of these issues.

I have never before had time to read many books. Now, at last, I have the time, and have traveled through so many adventures in real life and fiction that I can no longer feel comfortable sitting down without a book. Of course, the computer has been another wonderful awakening, and I cannot imagine who stocks up all those search engines with such a vast amount of information. It is so much quicker and more convenient than lifting down those heavy encyclopedias, where one often only finds a single entry, while a search engine might have 2,000 entries on the same topic and many related entries.

It has been a long sequence of colorful events that has taken me from the old woman squatting beside her bowls of peppers and onions in Colombo, to a meeting with the Emperor of Japan. I regret that I have failed to recall all the wonderful persons whom I have had the good fortune to know on the way. In school, there were many friendly colleagues, in university, there were other students who sharpened my wits. In my career, there were the unnamed students, technicians, and fellow scientists who either kindly or unkindly, brought about the need for me to rethink different situations. Mrs. Clare Spurgin, who gave me my presentation to Christ's Hospital, wrote a book late in her life called *My Journey*, which was printed privately. The book is largely a compendium of all the persons that she met in her lifetime. Her story is lost in the honors she bestowed on others. In my life, it is the system which I have tried to honor—the winds of change that allowed so much to happen.

In conclusion, but certainly not least, it seems to me that the world is short of affection for life, and long on dissatisfaction with living. There is only so much time in life and we never know how much that is, only that it ends. It would seem that cultivating one's affections for all aspects of life is a healthy enterprise.

But at my back I alwaies hear
Times winged Chariot hurrying near:
The Grave's a fine and private place,
But none I think do there embrace.

(Andrew Marvell [1621-1678] "To His Coy Mistress")

A Refit for the Future

ISBN 142511413-X